普通高等学校双语教学规划教材

Optical Sensing and Measurement

主编 徐 强

参编 代少玉 刘春波 马红玉 李平舟

西安电子科技大学出版社

内 容 简 介

本书主要介绍光学传感与测量的基本原理、基本概念和基本技术，可使读者了解光学传感与测量领域的新成果和新进展。本书内容主要包括光传输基本理论、光纤工作原理、常用激光器及工作原理、光度学、辐射度学、光辐射探测器、微弱光电信号检测等。书中提供了大量的光学传感与测量技术的应用实例，可为读者今后从事相关方面的研究和开发工作打下一定的基础。

本书着重基础知识的阐述，对内容深度作了适当考虑，可作为光学传感与测量工作人员的参考用书，也可作为高等院校相关专业的双语教材。

图书在版编目(CIP)数据

光学传感与测量= Optical Sensing and Measurement:英文 /徐强主编. －西安：西安电子科技大学出版社，2017.12
ISBN 978-7-5606-4565-0

Ⅰ.① 光… Ⅱ.① 徐… Ⅲ.① 光纤传感器－英文 ② 光学测量－英文
Ⅳ.① TP212.4 ② TB96

中国版本图书馆 CIP 数据核字(2017)第 189063 号

策　　划	云立实
责任编辑	卢　杨　云立实
出版发行	西安电子科技大学出版社(西安市太白南路2号)
电　　话	(029)88242885　88201467　　邮　编　710071
网　　址	www.xduph.com　　　电子邮箱　xdupfxb001@163.com
经　　销	新华书店
印刷单位	陕西天意印务有限责任公司
版　　次	2017年12月第1版　2017年12月第1次印刷
开　　本	787毫米×960毫米　1/16　印张　10.5
字　　数	208千字
印　　数	1～1000册
定　　价	22.00元

ISBN 978-7-5606-4565-0/TP
XDUP 4857001-1

＊＊＊如有印装问题可调换＊＊＊

前　言

光学传感与测量是近年来发展迅猛的综合性高新技术。从20世纪70年代后期起，随着半导体光电子器件和硅基光导纤维两大基础元件在原理和制造工艺上的突破，光子技术与电子技术开始结合并形成了具有强大生命力的信息光电子技术和产业。它包括信息传输（如光纤通信、空间和海底光通信等）、信息处理（如计算机光互连、光计算、光交换等）、信息获取（如光学传感和遥感、光纤传感等）、信息存储（如光盘、全息存储技术等）、信息显示（如大屏幕平板显示、激光打印和印刷等）。

本书较全面地讲述了光学传感与测量的基本原理、规律和方法，包括固体中光发射、光调制、光的传输和耦合、光的探测和接收及光信息处理、记录和显示等，并注意介绍该领域中一些新发展和新技术。本书一方面注重光电子技术的基础内容，力图体现光电子技术全貌，另一方面适当加入了近年一些相关领域的研究和应用成果，使其更符合光学传感与测量迅速发展的要求。

在本书编写过程中，编者得到了美国田纳西大学（The University of Tennessee）Marianne Breinig 教授的大力支持，在此表示由衷的感谢。

本书可作为高等院校电子信息、电子科学与技术、光信息科学与技术等专业的双语教学教材，亦可作为相关专业科研人员和工程技术人员的参考用书。

编者
2017 年 6 月

CONTENTS

Chapter 1 Nature of Light 1
  1.1 Light—Wave or Stream of Particles? 1
    1.1.1 What is a wave? 1
    1.1.2 Evidence for wave properties of light 2
    1.1.3 Evidence for light as a stream of particles 2
  1.2 Features of A Wave 3
  1.3 Electromagnetic Wave 4
    1.3.1 A periodic, sinusoidal wave 4
    1.3.2 Huygens' wavelets 7
  1.4 Maxwell's Equations 9
    1.4.1 Classical electrodynamics 13
    1.4.2 Properties of photons 16
  1.5 Reflection, Snell's Law 19
    1.5.1 Reflection 19
    1.5.2 Total internal reflection 21
  1.6 Fresnel Equations 22
  1.7 Fermat's Principle 29
    1.7.1 Fermat's principle yields Snell's law. 31
    1.7.2 The single slit 42
  1.8 Interference 49
    1.8.1 Constructive and destructive interference 50
    1.8.2 Young's double-slit interference experiment ... 51
  1.9 Diffraction 54
    1.9.1 Diffraction 54
    1.9.2 Diffraction by a single slit 55

    1.9.3 Diffraction grating ⋯⋯⋯⋯⋯⋯⋯⋯⋯⋯⋯⋯⋯⋯⋯⋯⋯⋯ 57

Chapter 2 **Polarization** ⋯⋯⋯⋯⋯⋯⋯⋯⋯⋯⋯⋯⋯⋯⋯⋯⋯⋯⋯⋯⋯⋯⋯⋯⋯⋯⋯⋯ 61
  2.1 Polarization of Light Waves ⋯⋯⋯⋯⋯⋯⋯⋯⋯⋯⋯⋯⋯⋯⋯⋯⋯⋯⋯⋯ 61
  2.2 Law of Malus ⋯⋯⋯⋯⋯⋯⋯⋯⋯⋯⋯⋯⋯⋯⋯⋯⋯⋯⋯⋯⋯⋯⋯⋯⋯⋯⋯ 64
  2.3 Polarization by Reflection and Brewster's Angle ⋯⋯⋯⋯⋯⋯⋯⋯ 66
  2.4 Brewster Windows in A Laser Cavity ⋯⋯⋯⋯⋯⋯⋯⋯⋯⋯⋯⋯⋯⋯ 68
  2.5 Polarization and Electromagnetic Effects ⋯⋯⋯⋯⋯⋯⋯⋯⋯⋯⋯⋯ 69

Chapter 3 **Fiber Optics** ⋯⋯⋯⋯⋯⋯⋯⋯⋯⋯⋯⋯⋯⋯⋯⋯⋯⋯⋯⋯⋯⋯⋯⋯⋯⋯⋯⋯ 79
  3.1 Geometrical Optics ⋯⋯⋯⋯⋯⋯⋯⋯⋯⋯⋯⋯⋯⋯⋯⋯⋯⋯⋯⋯⋯⋯⋯⋯ 79
  3.2 Wave Optics ⋯⋯⋯⋯⋯⋯⋯⋯⋯⋯⋯⋯⋯⋯⋯⋯⋯⋯⋯⋯⋯⋯⋯⋯⋯⋯⋯⋯ 81
  3.3 Optical Connectors ⋯⋯⋯⋯⋯⋯⋯⋯⋯⋯⋯⋯⋯⋯⋯⋯⋯⋯⋯⋯⋯⋯⋯⋯ 88
  3.4 Optical Fiber Measurements ⋯⋯⋯⋯⋯⋯⋯⋯⋯⋯⋯⋯⋯⋯⋯⋯⋯⋯⋯ 90
  3.5 Fiber-optic Communication ⋯⋯⋯⋯⋯⋯⋯⋯⋯⋯⋯⋯⋯⋯⋯⋯⋯⋯⋯⋯ 94
  3.6 Integrated Optics ⋯⋯⋯⋯⋯⋯⋯⋯⋯⋯⋯⋯⋯⋯⋯⋯⋯⋯⋯⋯⋯⋯⋯⋯⋯ 97

Chapter 4 **Stimulated Emission Devices Lasers** ⋯⋯⋯⋯⋯⋯⋯⋯⋯⋯⋯⋯⋯ 104
  4.1 Light Amplification, Resonators ⋯⋯⋯⋯⋯⋯⋯⋯⋯⋯⋯⋯⋯⋯⋯⋯ 104
  4.2 Types and Operating Principles ⋯⋯⋯⋯⋯⋯⋯⋯⋯⋯⋯⋯⋯⋯⋯⋯⋯ 115

Chapter 5 **Sources and Detectors** ⋯⋯⋯⋯⋯⋯⋯⋯⋯⋯⋯⋯⋯⋯⋯⋯⋯⋯⋯⋯⋯⋯ 122
  5.1 The Source of Light ⋯⋯⋯⋯⋯⋯⋯⋯⋯⋯⋯⋯⋯⋯⋯⋯⋯⋯⋯⋯⋯⋯⋯ 122
  5.2 Radiometric Quantities ⋯⋯⋯⋯⋯⋯⋯⋯⋯⋯⋯⋯⋯⋯⋯⋯⋯⋯⋯⋯⋯ 123
  5.3 Photometric Quantities ⋯⋯⋯⋯⋯⋯⋯⋯⋯⋯⋯⋯⋯⋯⋯⋯⋯⋯⋯⋯⋯ 124
  5.4 Radiation Laws ⋯⋯⋯⋯⋯⋯⋯⋯⋯⋯⋯⋯⋯⋯⋯⋯⋯⋯⋯⋯⋯⋯⋯⋯⋯ 129
  5.5 Optical Detectors ⋯⋯⋯⋯⋯⋯⋯⋯⋯⋯⋯⋯⋯⋯⋯⋯⋯⋯⋯⋯⋯⋯⋯⋯ 131
    5.5.1 Light detection ⋯⋯⋯⋯⋯⋯⋯⋯⋯⋯⋯⋯⋯⋯⋯⋯⋯⋯⋯⋯ 131
    5.5.2 Detector characteristics ⋯⋯⋯⋯⋯⋯⋯⋯⋯⋯⋯⋯⋯⋯⋯⋯ 134
    5.5.3 Noise considerations ⋯⋯⋯⋯⋯⋯⋯⋯⋯⋯⋯⋯⋯⋯⋯⋯⋯ 138
    5.5.4 Types of detectors ⋯⋯⋯⋯⋯⋯⋯⋯⋯⋯⋯⋯⋯⋯⋯⋯⋯⋯ 142
    5.5.5 Calibration ⋯⋯⋯⋯⋯⋯⋯⋯⋯⋯⋯⋯⋯⋯⋯⋯⋯⋯⋯⋯⋯⋯ 156
    5.5.6 Power supplies for optical detectors ⋯⋯⋯⋯⋯⋯⋯⋯ 158

References ⋯⋯⋯⋯⋯⋯⋯⋯⋯⋯⋯⋯⋯⋯⋯⋯⋯⋯⋯⋯⋯⋯⋯⋯⋯⋯⋯⋯⋯⋯⋯⋯⋯⋯⋯⋯⋯⋯ 162

Chapter 1 Nature of Light

1.1 Light—Wave or Stream of Particles?

Answer: Yes! As we'll see below, there is experimental evidence for both interpretations, although they seem contradictory.

1.1.1 What is a wave?

More familiar types of waves are sound, or waves on a surface of water. In both cases, there is a perturbation with a periodic spatial pattern which propagates, or travels in space. In the case of sound waves in air for example, the perturbed quantity is the pressure, which oscillates about the mean atmospheric pressure. In the case of waves on a water surface, the perturbed quantity is simply the height of the surface, which oscillates about its stationary level. Figure 1.1 shows an example of a wave, captured at a certain instant in time. It is simpler to visualize a wave by drawing the "wave fronts", which are usually taken to be the crests of the wave. In the case of Figure 1.1, the wave fronts are circular, as shown below the wave plot.

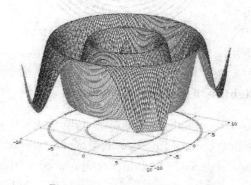

Figure 1.1 The wave fronts

Optical Sensing and Measurement

1.1.2 Evidence for wave properties of light

There are certain things that only waves can do, for example interfere. Ripples in a pond caused by two pebbles dropped at the same time exhibit this nicely: where two crests overlap, the waves reinforce each other, but where a crest and a trough coincide, the two waves actually cancel. This is illustrated in Figure 1.2. If light is a wave, two sources emitting waves in a synchronized fashion should produce a pattern of alternating bright and dark bands on a screen. Thomas Young tried the experiment in the early 1800's, and found the expected pattern. The wave model of light has one serious drawback, though: unlike other wave phenomena such as sound, or surface waves, it wasn't clear what the medium was that supported light waves. Giving it a name—the "luminiferous aether"—didn't help. James Clerk Maxwell's (1831 – 1879) theory of electromagnetism, however, showed that light was a wave in combined electric and magnetic fields, which, being force fields, didn't need a material medium.

Figure 1.2 Two sources emitting waves

1.1.3 Evidence for light as a stream of particles

One of the earliest proponents of the idea that light was a stream of particles was Isaac Newton himself. Although Young's findings and others seemed to disprove that theory entirely, surprisingly other experimental evidence appeared at the turn of the 20th century which could only be explained by the particle model of light! The photoelectric effect, where light striking a metal dislodges electrons from the metal atoms which can then flow

Chapter 1 Nature of Light

as a current earned Einstein the Nobel Prize for his explanation in terms of photons.

We are forced to accept that both interpretations of the phenomenon of light are true, although they appear to be contradictory. One interpretation or the other will serve better in a particular context. For our purposes, in understanding how optical instruments work, the wave theory of light is entirely adequate.

1.2 Features of A Wave

We'll consider the simple case of a sine wave in 1 dimension, as shown in Figure 1.3. The distance between successive wave fronts is the wavelength. As the wave propagates, let us assume in the positive x direction, any point on the wave pattern is displaced by dx in a time dt (see Figure 1.4). We can speak of the propagation speed of the wave

$$v = \frac{dx}{dt} \tag{1.1}$$

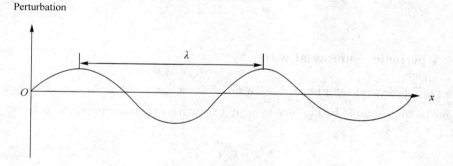

Figure 1.3　A sine wave in one dimension

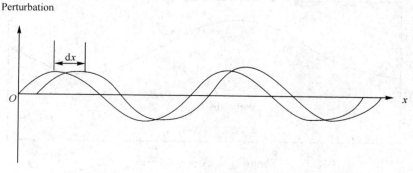

Figure 1.4　The propagation of the wave

Optical Sensing and Measurement

As the wave propagates, so do the wavefronts. A stationary observer in the path of the wave would see the perturbation oscillate in time, periodically in "cycles". The duration of each cycle is the period of the wave, and the number of cycles measured by the observer each second is the frequency. There is a simple relation between the wavelength λ, frequency f, and propagation speed v of a wave:

$$v = f \cdot \lambda \qquad (1.2)$$

Electromagnetic waves in vacuum always propagate with speed $c = 3.0 \times 10^8$ m/s. In principle, electromagnetic waves may have any wavelength, from zero to arbitrarily long. Only a very narrow range of wavelengths, approximately 400 – 700 nm, are visible to the human eye. We perceive wavelength as color; the longest visible wavelengths are red, and the shortest are violet. Longer than visible wavelengths are infrared, microwave, and radio. Shorter than visible wavelengths are ultraviolet, X-rays, and γ-rays.

1.3   Electromagnetic Wave

1.3.1   A periodic, sinusoidal wave

In a periodic wave (see Figure 1.5) a pulse travels a distance of one wavelength λ in a time equal to one period T. The speed v of the wave can be expressed in terms of these quantities.

$$v = \lambda / T = \lambda f \qquad (1.3)$$

This relationship holds true for any periodic wave.

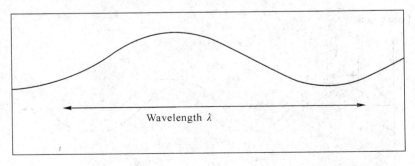

Figure 1.5   A periodic wave

If the individual atoms and molecules in the medium move with simple harmonic motion, the resulting periodic wave has a sinusoidal form. We call it a **harmonic wave** or a **sinusoidal wave**.

If the displacement of the individual atoms or molecules is perpendicular to the direction the wave is traveling, the wave is called a **transverse wave**(see Figure 1.6(a)).

If the displacement is parallel to the direction of travel the wave is called a **longitudinal wave** or a **compression wave**(see Figure 1.6(b)).

(a)　　A transverse wave　　　　　　　　(b)　　A longitudinal wave

Figure 1.6　A harmonic wave

Transverse waves can occur only in solids, whereas longitudinal waves can travel in solids, fluids, and gases. Transverse motion requires that each particle drag with it adjacent particles to which it is tightly bound. In a fluid this is impossible, because adjacent particles can easily slide past each other. Longitudinal motion only requires that each particle push on its neighbors, which can easily happen in a fluid or gas. The fact that longitudinal waves originating in an earthquake pass through the center of the earth while transverse waves do not is one of the reasons the earth is believed to have a liquid outer core.

Consider a transverse harmonic wave traveling in the positive x-direction. The displacement y of a particle in the medium is given as a function of x and t by

$$y(x, t) = A\sin(kx - \omega t + \phi) \tag{1.4}$$

Here k is the **wavenumber**, $k = 2\pi/\lambda$, and $\omega = 2\pi/T = 2\pi f$ is the **angular frequency** of the wave. ϕ is called the **phase constant**. At a fixed time t the displacement varies as a function of x as $A\sin(\theta) = A\sin(kx) = A\sin[(2\pi/\lambda)x]$. The phase constant ϕ is determined by the initial conditions of the motion. If at $t=0$ and $x=0$ the displacement is zero then $\phi=0$ or π. If at $t=0$ and $x=0$ the displacement has its maximum value, then $\phi=\pi/2$. The quantity $kx - \omega t + \phi$ is called the **phase**. Figure 1.7 shows the phase of wave.

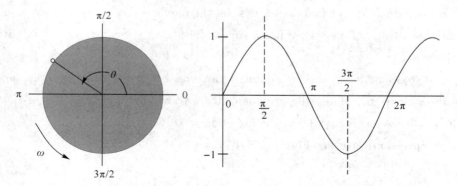

Figure 1.7 The phase of wave

At a fixed position x the displacement varies as a function of time as $A\sin(\theta)=A\sin(\omega t)$ with a convenient choice of origin. At a fixed time the displacement varies as a function of position as $A\sin(\theta)=A\sin(kx)$ with a convenient choice of origin Figure 1.8 shows the displacement of a wave varies as a function of time and position.

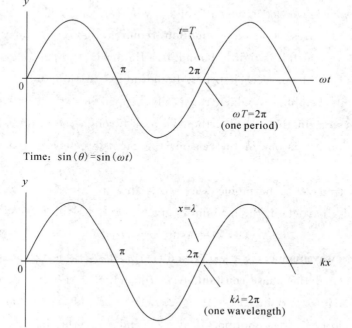

Figure 1.8 The displacement of a wave varies as a function of time and position

Chapter 1 Nature of Light

For a harmonic wave $y(x, t) = A\cos(kx - \omega t + \phi)$ traveling in the **positive x-direction** we may write

$$y(x, t) = A\sin\left[\left(\frac{2\pi}{\lambda}\right)x - (2\pi f)t + \phi\right] = A\sin\left[\left(\frac{2\pi}{\lambda}\right)(x - \lambda f t) + \phi\right] \quad (1.5)$$

or, using $\lambda f = v$ and $\pi/\lambda = k$,

$$y(x, t) = A\sin[k(x - vt) + \phi] \quad (1.6)$$

For a harmonic wave $y(x, t) = A\sin(kx + \omega t + \phi)$ traveling in the **negative x-direction** we have

$$y(x, t) = A\sin(kx + \omega t + \phi) = A\sin[k(x + vt) + \phi] \quad (1.7)$$

The **amplitude** A of a wave is the maximum displacement of the individual particles from their equilibrium position. The **energy** E carried by a wave is proportional to the square of its amplitude.

$$E \propto A^2 \quad (1.8)$$

The **power** P delivered by the wave if it is absorbed is proportional to the square of its amplitude times its speed.

$$P \propto A^2 v$$

Two or more waves traveling in the same medium travel independently and can pass through each other. In regions where they overlap we only observe a single disturbance. We observe **interference**. When two or more waves interfere, **the resulting displacement is equal to the vector sum of the individual displacements**. If the displacements have the same vector directions and two waves with equal amplitudes overlap **in phase**, i.e. if crest meets crest and trough meets trough, then we observe a resultant wave with twice the amplitude. We have **constructive interference**. If the two overlapping waves, however, are completely **out of phase**, i.e. if crest meets trough, then the two waves cancel each other out completely. We have **destructive interference**.

When the medium through which a wave travels abruptly changes, the wave may be partially or totally reflected. If a periodic wave is reflected, then the incident wave and the reflected wave travel in the same medium in opposite directions and the reflected wave interferes with the incoming wave. The resulting pattern can become very complex.

1.3.2 Huygens' wavelets

Long before people understood the electromagnetic character of light, Christian Huygens, a 17th-century scientist, came up with a technique for propagating waves from one position to another, determining, in effect, the shapes of the developing wave fronts. This technique is basic to a quantitative study of interference and diffraction, so we cover it here briefly. Huygens claimed that:

Every point on a known wave front in a given medium can be treated as a point source of secondary wavelets (spherical waves "bubbling" out of the point, so to speak) which spread out in all directions with a wave speed characteristic of that medium. The developing wave front at any subsequent time is the envelope of these advancing spherical wavelets.

Figure 1.9 shows how Huygens' principle is used to demonstrate the propagation of successive (a) *plane* wave fronts and (b) *spherical* wave fronts. Huygens' technique involves the use of a series of points $P_1, \ldots, P_8$, for example, on a given wave front defined at a time $t=0$. From these points, as many as one wishes, actually, spherical wavelets are assumed to emerge, as shown in Figures 1.9(a) and 1.9(b). Radiating outward from each of the P-points, with a speed v, the series of secondary wavelets of radius $r=vt$ defines a new wave front at some time t later. In Figure 1.9(a) the new wave front is drawn as an *envelope tangent* to the secondary wavelets at a distance $r=vt$ from the initial plane wave front. It is, of course, another *plane* wave front. In Figure 1.9(b), the new wave front at time t is drawn as an *envelope tangent* to the secondary wavelets at a distance $r=vt$ from the initial spherical wave front. It is an advancing *spherical* wave front.

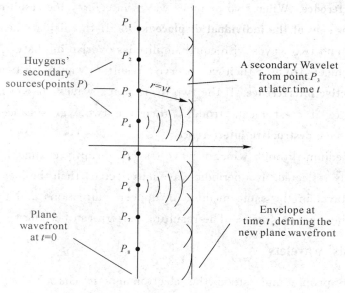

(a) Plane waves

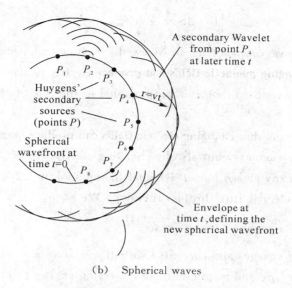

(b) Spherical waves

Figure 1.9 Huygens' principle applied to the propagation of plane and spherical wave fronts

While there seems to be no physical basis for the existence of Huygens' "secondary" point sources, Huygens' technique has enjoyed extensive use, since it does predict accurately, with waves, not rays, both the *law of reflection and Snell's law of refraction*. In addition, Huygens' principle forms the basis for calculating, for example, the diffraction pattern formed with multiple slits. We shall soon make use of Huygens' secondary sources when we set up the problem for diffraction from a single slit.

1.4 Maxwell's Equations

Electromagnetic (EM) waves can transport energy across empty space. The equations of electrodynamics are Maxwell's equations.

$$\oint_A E \cdot dA = \frac{Q_{\text{inside}}}{\varepsilon_0} \tag{1.9a}$$

$$\oint_\Gamma E \cdot dS = -\frac{\partial \Phi_B}{\partial t} \tag{1.9b}$$

$$\oint_A B \cdot dA = 0 \tag{1.9c}$$

Optical Sensing and Measurement

$$\oint_\Gamma B \cdot dS = \mu_0 I_{through\Gamma} + \frac{1}{C^2} \frac{\partial \Phi_E}{\partial t} \tag{1.9d}$$

Electromagnetic waves are solutions to Maxwell's equations. Equation (1.9b) (Faraday's law) tells us that **changing magnetic fields can produce electric fields**. The circulation of the electric field around any closed loop Γ is proportional to the rate of change of the magnetic flux through the loop.

Equation 1.9 tells us that **changing electric fields can produce magnetic fields**. Magnetic fields are produced by currents, but also by changing electric fields. The circulation of the magnetic field around any closed loop Γ is equal to the sum of $\mu_0 I_{through}$ and $1/c^2$ times the rate of change of the electric flux through the loop. We have

$$\mu_0 \varepsilon_0 = \frac{1}{c^2} \tag{1.10}$$

In a space free of charges and currents, we still can have electric and magnetic fields. They are changing electric and magnetic fields, carrying energy through space. EM waves require no medium; they can travel through empty space. Sinusoidal plane waves are one type of electromagnetic waves. Not all EM waves are sinusoidal plane waves, but all electromagnetic waves can be viewed as a linear superposition of sinusoidal plane waves traveling in arbitrary directions. A plane EM wave traveling in the x-direction is of the form

$$E(x, t) = E_{max} \cos(kx - \omega t + \phi)$$
$$B(x, t) = B_{max} \cos(kx - \omega t + \phi) \tag{1.11}$$

E is the electric field vector and B is the magnetic field vector of the EM wave (see Figure 1.10). For electromagnetic waves E and B are always perpendicular to each other, and perpendicular to the direction of propagation. The direction of propagation is the direction of $E \times B$.

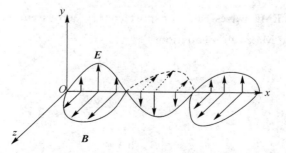

Figure 1.10 The magnetic field vector of the EM wave

If, for a wave traveling in the x-direction $\boldsymbol{E} = E\boldsymbol{j}$, then $\boldsymbol{B} = B\boldsymbol{k}$ and $\boldsymbol{j} \times \boldsymbol{k} = \boldsymbol{i}$. **Electromagnetic waves are transverse waves.**

The wave vector $\boldsymbol{k}$ points into the direction of propagation, and its magnitude $k = 2\pi/\lambda$, where λ is the wavelength of the wave. The frequency f of the wave is $f = \omega/2\pi$. ω is the angular frequency. The speed of any sinusoidal wave is the product of its wavelength and frequency.

$$v = \lambda f = \frac{\omega}{k} \tag{1.12}$$

Maxwell's equations require that $v = c = 3 \times 10^8$ m/s for any electromagnetic wave in free space. **The speed of any electromagnetic waves in free space is the speed of light c.** Periodic electromagnetic waves in free space can have any wavelength λ or frequency f as long as $\lambda f = c$. When an electromagnetic wave travels through free space, Maxwell's equations require that **at every instant and at any point the ratio of the electric to the magnetic field in SI units is equal to the speed of light**, $\boldsymbol{E}/\boldsymbol{B} = c$.

When electromagnetic waves travel through a medium, the speed of the waves in the medium is

$$v = \frac{c}{n(\lambda_{\text{free}})} \tag{1.13}$$

where $n(\lambda_{\text{free}})$ is the index of refraction of the medium. The index of refraction n is a properties of the medium, and it depends on the wavelength λ_{free} of the EM wave. If the medium absorbs some of the energy transported by the wave, then $n(\lambda_{\text{free}})$ is a complex number. For air n is nearly equal to 1 for all wavelengths.

When an EM wave travels from one medium with index of refraction n_1 into another medium with a different index of refraction n_2, then its **frequency remains the same**, but its speed and wavelength change.

Electromagnetic waves are categorized according to their frequency f or, equivalently, according to their wavelength $\lambda = c/f$. Visible light has a wavelength range from $\sim$400 nm to $\sim$700 nm. Violet light has a wavelength of $\sim$400 nm, and a frequency of $\sim 7.5 \times 10^{14}$ Hz. Red light has a wavelength of $\sim$700 nm, and a frequency of $\sim 4.3 \times 10^{14}$ Hz(see Figure 1.11).

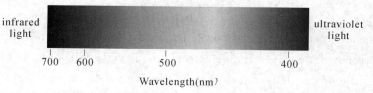

Figure 1.11 Wavelengths of Electromagnetic waves

Optical Sensing and Measurement

Visible light makes up just a small part of the full **electromagnetic spectrum**. Electromagnetic waves with shorter wavelengths and higher frequencies include ultraviolet light, x-rays, and gamma rays. Electromagnetic waves with longer wavelengths and lower frequencies include infrared light, microwaves, and radio and television waves.

Type of Radiation	Frequency Range/Hz	Wavelength Range
gamma-rays	$10^{20} - 10^{24}$	$< 10^{-12}$ m
x-rays	$10^{17} - 10^{20}$	1 nm – 1 pm
ultraviolet	$10^{15} - 10^{17}$	400 nm – 1 nm
visible	$4 - 7.5 \times 10^{14}$	750 nm – 400 nm
near-infrared	$1 \times 10^{14} - 4 \times 10^{14}$	2.5 μm – 750 nm
infrared	$10^{13} - 10^{14}$	25 μm – 2.5 μm
microwaves	$3 \times 10^{11} - 10^{13}$	1 mm – 25 μm
radio waves	$< 3 \times 10^{11}$	> 1 mm

Electromagnetic waves transport energy through space. In free space this energy is transported by the wave with speed c. The magnitude of the **energy flux S** is the amount of energy that crosses a unit area perpendicular to the direction of propagation of the wave per unit time. It is given by

$$S = \frac{EB}{(\mu_0)} = \frac{E^2}{(\mu_0 c)} \tag{1.14}$$

since for electromagnetic waves $B = E/c$. The units of S are J/(m² · s). μ_0 is a constant called the permeability of free space, $\mu_0 = 4\pi \times 10^{-7}$ N/A².

The **Poynting vector** is the energy flux vector. It is named after John Henry Poynting. Its direction is the direction of propagation of the wave, i.e. the direction in which the energy is transported.

$$S = \left(\frac{1}{\mu_0}\right) E \times B \tag{1.15}$$

Energy per unit area per unit time is power per unit area. S represents the power per unit area in an electromagnetic wave. If an electromagnetic wave falls onto an area A where it is absorbed, then the power delivered to that area is $P = S \times A$.

The time average of the magnitude of the Poynting vector, $\langle S \rangle$, is called the *irradiance* or intensity. The irradiance is the average energy per unit area per unit time. $\langle S \rangle = \langle E^2 \rangle / (\mu_0 c) = E_{max}^2 / (2\mu_0 c)$.

Electromagnetic waves transport energy. **EM wave also transport momentum.** The momentum flux is S/c. S/c is the amount of momentum that crosses a unit area perpendicular to the direction of propagation of the wave per unit time. If an electromagnetic wave falls onto an area A where it is absorbed, the momentum delivered to that area in a direction perpendicular to the area per unit time is $dp_\perp/dt = (1/c) S \cdot A$.

The momentum of the object absorbing the radiation therefore changes. The rate of change is $dp_\perp/dt = (1/c) SA_\perp$, where $A_\perp$ is the cross-sectional area of the object perpendicular to the direction of propagation of the electromagnetic wave. The momentum of an object changes if a force is acting on it.

$$F_\perp = \frac{dp_\perp}{dt} = \left(\frac{1}{c}\right) SA_\perp \tag{1.16}$$

is the force exerted by the radiation on the object that is absorbing the radiation. Dividing both sides of this equation by $A_\perp$, we find the **radiation pressure** (force per unit area) $P = (1/c) S$. If the radiation is reflected instead of absorbed, then its momentum changes direction. The radiation pressure on an object that reflects the radiation is therefore twice the radiation pressure on an object that absorbs the radiation.

1.4.1 Classical electrodynamics

▶ **A charged particle** produces an electric field. This electric field exerts a force on other charged particles. Positive charges accelerate in the direction of the field and negative charges accelerate in a direction opposite to the direction of the field.

▶ **A moving charged particle** produces a magnetic field. This magnetic field exerts a force on other moving charges. The force on these charges is always perpendicular to the direction of their velocity and therefore only changes the direction of the velocity, not the speed.

▶ **An accelerating charged particle** produces an electromagnetic wave. Electromagnetic waves are electric and magnetic fields traveling through empty space with the speed of light c. A charged particle oscillating about an equilibrium position is an accelerating charged particle. If its frequency of oscillation is f, then it produces an electromagnetic wave with frequency f. The wavelength λ of this wave is given by $\lambda = c/f$. Electromagnetic waves transport energy through space. This energy can be delivered to charged particles a large distance away from the source. Figure 1.12 shows the electric field of the electromagnetic wave produced by the accelerating charge.

Assume a charge q is accelerating. It therefore produces electromagnetic radiation. At some position r in space and at some time t, the electric field of the electromagnetic wave, (i. e. the radiation field,) produced by the accelerating charge is given by

$$E(r, t) = -\left[\frac{1}{4\pi\varepsilon_0}\right]_{SI} \frac{q}{c^2 r''} a_\perp \left(t - \frac{r''}{c}\right) \quad (1.17)$$

where

$$r'' = r - r'\left(t - \frac{|r - r'|}{c}\right) \quad (1.18)$$

Let us analyze this expression. The magnitude of the electric field is proportional to the charge q. The bigger the accelerating charge, the bigger is the magnitude of the field. The magnitude decreases as the inverse of the distance r'', which is the distance between the accelerating charge and the position where the field is observed. But it is not the distance at the time the field is observed, but the distance at some earlier time, called the **retarded time**, when the radiation field was produced. Since electromagnetic waves travel with speed c, it takes them a time interval $\Delta t = \Delta r/c$ to travel a distance Δr. The electric field is also proportional to the acceleration of the charge. The larger the acceleration, the larger is the field. The directional aspects are given by $E(r, t)$, $a_\perp$. The direction of the electric field is perpendicular to the line of sight between r and the retarded position of the charge and its magnitude is proportional to the component of the acceleration perpendicular to this line of sight. The figure below (see Figure 1.12) illustrates that point. The electric field is zero along a line of sight in the direction of the acceleration, largest along a line of sight perpendicular to the direction of the acceleration, and always perpendicular to the line of sight.

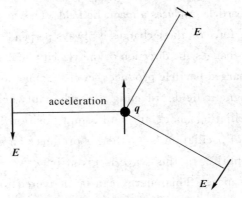

Figure 1.12 The electric field of the electromagnetic wave produced by the accelerating charge

The magnetic field is perpendicular to the electric field and to the direction of propagation r'', and its magnitude is $B=E/c$. This can be written as

$$B=\frac{\hat{r}''}{c}\times \boldsymbol{E} \tag{1.19}$$

Since the radiation decreases as the inverse of the distance r'', the irradiance decreases as the inverse of r''^2. This is the well-known **inverse square law.**

The power radiated away by the non-relativistically moving, accelerating charge is

$$p=\oint \boldsymbol{S}\cdot\hat{n}\mathrm{d}A = \frac{2}{3}\frac{e^2 a^2}{c^3}, \text{ with } e^2=\frac{q^2}{4\pi\varepsilon_0}$$

This is called the **Lamor formula.**

Polarization is a phenomenon peculiar to transverse waves. Longitudinal waves such as sound cannot be polarized. Light and other electromagnetic waves are transverse waves made up of mutually perpendicular, fluctuating electric and magnetic fields. In the diagram below an EM wave is propagating in the x-direction, the electric field oscillates in the xy plane, and the magnetic field oscillates in the xz plane. A line traces out the electric field vector as the wave propagates. Figure 1.13 shows a polarized electromagnetic wave.

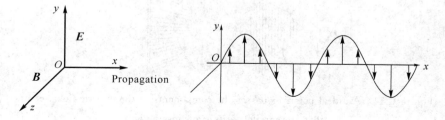

Figure 1.13 A polarized electromagnetic wave

An unpolarized electromagnetic wave traveling in the x-direction is a superposition of many waves. For each of these waves the electric field vector is perpendicular to the x-axis, but the angle it makes with the y-axis is different for different waves. For a polarized electromagnetic wave traveling in the x-direction, the angle the electric field makes with the y-axis is unique. A polarizer is a material that passes only EM waves for which the electric field vector is confined to a single plane that contains to the direction of motion.

An ideal polarizer passes the components of the electric field vectors that are parallel to its transmission axis. If E_0 is the incident field vector and the angle between E_0 and the transmission axis is θ, then the magnitude of transmitted field vector is $E_0\cos\theta$ and its

Optical Sensing and Measurement

direction is the direction of the transmission axis(see Figure 1.14). The irradiance I of an electromagnetic wave is proportional to the square of the magnitude of the electric field vector. We therefore have

$$I_{\text{transmitted}} = I_0 \cos^2\theta \tag{1.20}$$

If $\theta = 90°$ the transmitted intensity is zero.

All EM waves transport energy across space. The irradiance (energy per unit area and unit time) is proportional to the square of the average amplitude of the electric field of the EM wave. This energy, however, arrives at a receiver not continuously but in discrete units called **photons**. The energy transported by an electromagnetic wave is not continuously distributed over the wave front. It is transported in discrete packages. In addition to its wave properties, light also has particle nature. **Photons are the particles of light.**

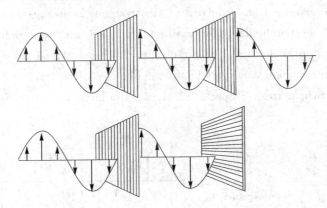

Figure 1.14 An ideal polarizer passes the components of the electric field vectors that are parallel to its transmission axis

1.4.2 Properties of photons

▶ Photons always move with the speed of light.
▶ Photons are electrically neutral.
▶ Photons have no mass, but they have energy $E = hf = hc/\lambda$. Here $h = 6.626 \times 10^{-34}$ J·s is a universal constant called **Planck's constant**. The energy of each photon is inversely proportional to the wavelength of the associated EM wave. The shorter the wavelength, the more energetic is the photon, the longer the wavelength, the less energetic is the photon.
▶ A laser beam and a microwave beam can carry the same amount of energy. In this

case the laser beam contains a smaller number of photons, but each photon in the laser beam has a higher energy than the photons in the microwave beam.

▶ Photons can be created and destroyed. When a source emits EM waves, photons are created. When photons encounter matter, they may be absorbed and transfer their energy to the atoms and molecules. Creation and destruction of photons must conserve energy and momentum. The magnitude of the momentum of a photon is $p=hf/c$, and the direction of the momentum is the direction of propagation of the EM wave.

The wave nature and the particle nature of light are complementary properties. Experiments probing the propagation of light through a medium and around obstacles reveal the wave nature of light. Experiments probing energy and momentum conservation when light interacts with atoms and molecules reveal the particle nature of light. No single experiment has ever revealed both the wave and particle nature simultaneously.

Figure 1.15 Albert Einstein

In 1905, Albert **Einstein** (see Figure 1.15) used the discrete nature of light to explain the **photoelectric effect**. To demonstrate this effect light is shone on a metal surface. If the frequency of the light is higher than the **cutoff frequency** f_c, then electrons are released. No photoelectric electrons are emitted if the frequency of the light falls below this cutoff frequency f_c. For many metal surfaces the frequency of blue light is greater than f_c and the frequency of red light is less than f_c. If red light is shone on the surface, no electrons are emitted, no matter what the intensity of the light (within limits, using normal light sources). If blue light is shone on the surface, electrons are emitted. The number of emitted electrons depends on the intensity of the light. But even if the intensity is reduced to a very low value, electrons are still emitted, albeit at a very low rate.

The photoelectric effect cannot be understood within the wave picture of light. To eject an electron from a metal surface a certain amount of energy ϕ, called the work function of the metal, must be supplied to this electron. In the wave picture the energy of the light beam does not depend on the frequency, but only on the intensity. Einstein explained the photoelectric effect by postulating that an electron can only receive the large amount of energy necessary to escape the metal from the EM wave by absorbing a single photon. If this photon has enough energy, the electron is freed. Excess energy appears as kinetic energy of the electron. The maximum kinetic energy of the electron is given by $E=hf-\phi$. If the photon does not have enough energy, then the electron cannot escape the metal.

Electromagnetic waves with wavelengths λ in the range of $\sim$400 nm to $\sim$750 nm are called **visible light**. We see light because it stimulates the cells in our eyes. Because our eyes are able to distinguish between different wavelength of light we perceive color. If the light reaching our eyes contains a broad mixture of wavelength, we interpret it as **white light**. Because light is an EM wave, it exhibits several behaviors characteristic of waves such as **reflection**, **refraction** and **diffraction** and **interference**.

A scheme for thinking about the nature of wave propagation is called Huygen's **principle**(see Figure 1.16). **Huygen's principle** is a geometrical construction that tells us how wavefronts will move, but not why they will move that way. Each point on a wave front is considered to be a point source for the production of new waves. In three dimensions, these new waves are spherical waves called **wavelets**, that propagate outward with the characteristic speed of the wave. The wavelets emitted by all points on the wave front interfere with each other to produce the traveling wave. When studying the propagation of light, we can replace any wave front by a collection of sources distributed uniformly over the wave front, radiating in phase.

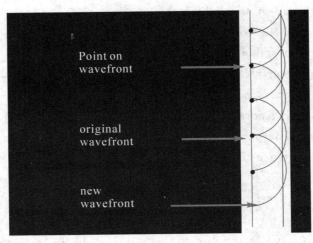

Figure 1.16 Huygen's principle

When light passes through a small opening, comparable in size to the wavelength of the light, in an otherwise opaque obstacle, the wave front on the other side of the opening resembles the wave front shown below.

The light spreads around the edges of the obstacle. This is the phenomenon of **diffraction** (see Figure 1.17).

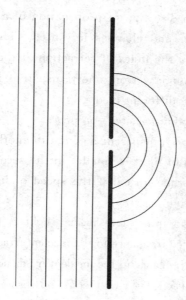

Figure 1.17   The phenomenon of diffraction

Light rays are orthogonal trajectories to the wave front of an EM wave with frequency in the visible region. They are lines normal to the wave front at every point of intersection. In an isotropic medium, light rays are parallel to the wave vector k.

1.5   Reflection, Snell's law

1.5.1   Reflection

Reflection is the abrupt change in the direction of propagation of a wave that strikes the boundary between two different media. At least some part of the incoming wave remains in the same medium. If an incoming light ray makes an angle θ_i with the normal of a plane tangent to the boundary, then the reflected ray makes an angle θ_r with this normal and lies in the same plane as the incident ray and the normal.

Law of reflection: $|\theta_r| = |\theta_i|$

The **reflectivity** of a surface material is the fraction of energy of the incoming wave that is reflected by it. The reflectivity of a mirror is close to 1.

Refraction is the change in direction of propagation of a wave when the wave passes from one medium into another and changes its speed. The speed of light in a given substance is $v = c/n$, where n is the **index of refraction** of the substance. Light waves are refracted when crossing the boundary from one transparent medium into another because the speed of light is different in different media.

Snell's law, or **the law of refraction**: $n_1 \sin\theta_1 = n_2 \sin\theta_2$.

When light passes from one transparent medium to another, the rays are bent toward the surface normal if the speed of light is smaller in the second medium than in the first. The rays are bent away from this normal if the speed of light in the second medium is greater than in the first (see Figure 1.18).

At a boundary between two transparent media, light is partially reflected and partially refracted. The ratio of the reflected irradiance to the incident irradiance is called the reflectance R and the ratio of the transmitted irradiance to the incident irradiance is called the transmittance T. Energy conservation requires that $R + T = 1$ (if there is no absorption).

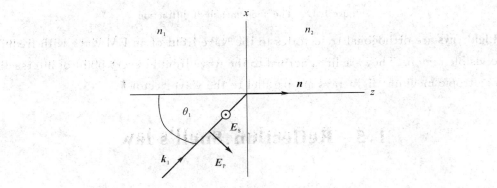

Figure 1.18 Light passes from one transparent medium to another

R and T depend on the indices of refraction of the two media n_1 and n_2, the angle of incidence θ_1, and the polarization of the incident light. We distinguish between p-polarization and s-polarization. Let the **plane of incidence** contain the normal to the boundary and the incident wave vector $\boldsymbol{k}_1$. The electric field vector $\boldsymbol{E}_1$ is perpendicular to $\boldsymbol{k}_1$. If we choose our coordinate system as shown below, then plane of incidence is the xz-plane and $\boldsymbol{E}_1$ may be written as $\boldsymbol{E}_1 = \boldsymbol{E}_p + \boldsymbol{E}_s$. $\boldsymbol{E}_p$ lies in the xz-plane and $\boldsymbol{E}_s$ is perpendicular

Chapter 1 Nature of Light

to the xz-plane, i.e. it points in the $\pm y$-direction. The electric field of the incident light is a linear superposition of p- and s-polarized fields.

For p-polarized light we have $R=|r_{12p}|^2$, where r_{12p} is the Fresnel reflection coefficient for p-polarization. We have

$$r_{12p}=\frac{\tan(\theta_1-\theta_2)}{\tan(\theta_1+\theta_2)} \tag{1.21}$$

For s-polarized light we have $R=|r_{12s}|^2$, where r_{12s} is the Fresnel reflection coefficient for s-polarization. We have

$$r_{12s}=\frac{\sin(\theta_1-\theta_2)}{\sin(\theta_1+\theta_2)} \tag{1.22}$$

For a graph of the reflectance R for s-and p-polarized light as a function of n_1, n_2, and θ_1.

If $\theta_1+\theta_2=\pi/2$, then $\tan(\theta_1+\theta_2)=\infty$ and $r_{12p}=0$. If light is reflected, it will have s-polarization. The incident angle at which this happens is called the Brewster angle θ_B. We then have

$$n_1\sin\theta_B=n_2\sin(\frac{\pi}{2}-\theta_B)=n_2\cos\theta_B$$

$$\tan\theta_B=\frac{n_2}{n_1} \tag{1.23}$$

Polarized light can thus be obtained via reflection.

1.5.2 Total internal reflection

Total internal reflection occurs only if light travels from a medium of high index of refraction to a medium of low index of refraction. Let light travel from medium 1 into medium 2 and let $n_1>n_2$. Then the critical angle θ_c is given by

$$\sin\theta_c=\frac{n_2}{n_1} \tag{1.24}$$

For angles greater than the critical angle the incident light is totally reflected, obeying the law of reflection.

The velocity of light in a material, and hence the index of refraction of the material, depends on the wavelength of the light. The index listed in tables is either an average index, or it is the index for one particular wavelength. Since the refractive index depends on the wavelength of the light, light waves with different wavelengths and therefore different colors are refracted through different angles. This is called **dispersion**, because white light is dispersed into its component colors while traveling through the material. In general, the index of refraction n varies inversely with wavelength. It is greater for shorter wavelengths.

1.6 Fresnel Equations

The **Fresnel equations**, deduced by Augustin-Jean Fresnel (see Figure 1.19), describe the behaviour of light when moving between media of differing refractive indices. The reflection of light that the equations predict is known as **Fresnel reflection** (see Figure 1.20).

Figure 1.19   Augustin Fresnel

(1788—1827)

When light moves from a medium of a given refractive index n_1 into a second medium with refractive index n_2, both reflection and refraction of the light may occur.

A light ray striking the interface between two media is split into two—a reflected part and a refracted part.

So far we have been able to deduce the laws of reflection and refraction using Huygens principle and Fermat's principle. Each gives a distinctive and valuable point of view. However, there is another, more powerful approach to the problem. It is provided by the electromagnetic theory of light.

In the diagram, an incident light ray strikes on the interface between two media of refractive indexes n_i and n_t. Part of the ray is reflected and part refracted. The angles that the incident, reflected and refracted rays make to the normal of the interface are given as

θ_i, θ_r and θ_t, respectively. The relationship between these angles is given by the law of reflection and Snell's law.

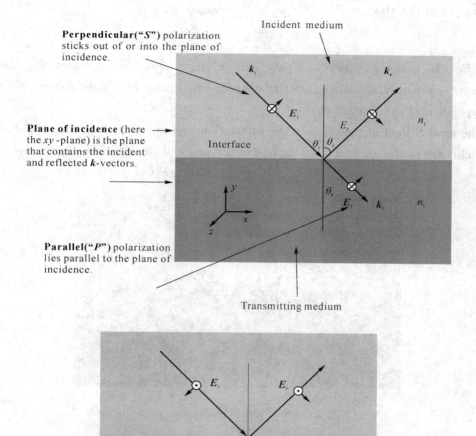

Figure 1.20 Fresnel reflection

We would like to compute the fraction of a light wave reflected and transmitted by a flat interface between two media with different refractive indices:
for the perpendicular polarization

$$r_\perp = \frac{E_{0r}}{E_{0i}}, \quad t_\perp = \frac{E_{0t}}{E_{0i}} \qquad (1.25)$$

for the parallel polarization

$$r_\parallel = \frac{E_{0r}}{E_{0i}}, \quad t_\parallel = \frac{E_{0t}}{E_{0i}} \qquad (1.26)$$

where E_{0i}, E_{0r}, and E_{0t} are the field complex amplitudes. We consider the boundary conditions at the interface for the electric and magnetic fields of the light waves. We will do the perpendicular polarization first.

The total E-field in the plane of the interface is continuous (see Figure 1.21). Here, all E-fields are in the z-direction, which is in the plane of the interface (xz), so:

$$E_i(x, y=0, z, t) + E_r(x, y=0, z, t) = E_t(x, y=0, z, t) \qquad (1.27)$$

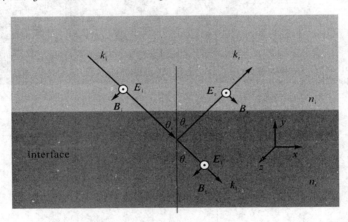

Figure 1.21 The total E-field in the plane of the interface

In other words:
The total B-field in the plane of the interface is continuous. Here, all B-fields are in the xy-plane, so we take the x-components:

$$-B_i(x, y=0, z, t)\cos(\theta_i) + B_r(x, y=0, z, t)\cos(\theta_r) = -B_t(x, y=0, z, t)\cos(\theta_t)$$
$$(1.28)$$

* It's really the tangential B/μ, but we're using $\mu = \mu_0$.

Canceling the rapidly varying parts of the light wave and keeping only the complex amplitudes:

$$E_{0i} + E_{0r} = E_{0t}$$
$$-B_{0i}\cos\theta_i + B_{0r}\cos\theta_r = -B_{0t}\cos\theta_t \qquad (1.29)$$

But $B = E/(c_0/n) = nE/c_0$ and $\theta_r = \theta_i$:

$$n_i(E_{0r} - E_{0i})\cos\theta_i = -n_t E_{0t}\cos\theta_t \tag{1.30}$$

Substituting for E_{0t} using $E_{0i} + E_{0r} = E_{0t}$:

$$n_i(E_{0r} - E_{0i})\cos\theta_i = -n_t(E_{0r} + E_{0i})\cos\theta_t \tag{1.31}$$

Rearranging $n_i(E_{0r} - E_{0i})\cos\theta_i = -n_t(E_{0r} + E_{0i})\cos\theta_t$ yields:

$$E_{0r}(n_i\cos\theta_i + n_t\cos\theta_t) = E_{0i}(n_i\cos\theta_i - n_t\cos\theta_t) \tag{1.32}$$

Solving for E_{0r}/E_{0i} yield the reflection coefficient:

$$r_\perp = \frac{E_{0r}}{E_{0i}} = \frac{n_i\cos\theta_i - n_t\cos\theta_t}{n_i\cos\theta_i + n_t\cos\theta_t} \tag{1.33}$$

Analogously, the transmission coefficient, E_{0t}/E_{0i} is

$$t_\perp = \frac{E_{0t}}{E_{0i}} = \frac{2n_i\cos\theta_i}{n_i\cos\theta_i + n_t\cos\theta_t} \tag{1.34}$$

These equations are called the Fresnel Equations for perpendicularly polarized light.

Note that the reflected magnetic field must point into the screen to achieve $\boldsymbol{E} \times \boldsymbol{B} \propto \boldsymbol{k}$. The x means "into the screen" (see Figure 1.22).

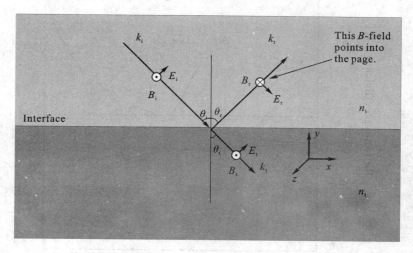

Figure 1.22 Beam geometry for light with its electric field **parallel** to the plane of incidence

For parallel polarized light,

$$B_{0i} - B_{0r} = B_{0t}$$

and

$$E_{0i}\cos\theta_i + E_{0r}\cos\theta_r = E_{0t}\cos\theta_t$$

Solving for E_{0r}/E_{0i} yields the reflection coefficient, $r^{\|}$:

$$r_{\|} = \frac{E_{0r}}{E_{0i}} = \frac{n_i \cos\theta_t - n_t \cos\theta_i}{n_i \cos\theta_t + n_t \cos\theta_i} \quad (1.35)$$

Analogously, the transmission coefficient, $t_{\|} = \dfrac{E_{0t}}{E_{0i}}$, is

$$t_{\|} = \frac{E_{0t}}{E_{0i}} = \frac{2 n_i \cos\theta_i}{n_i \cos\theta_t + n_t \cos\theta_i} \quad (1.36)$$

These equations are called the Fresnel Equations for **parallel** polarized light.

$$n_{air} \approx 1 < n_{glass} \approx 1.5$$

Total reflection at $\theta = 90°$ for both polarizations. Zero reflection for parallel polarization at **Brewster's angle** (56.3° for these values of n_i and n_t). Figure 1.23 shows reflection coefficients for an air-to-glass interface.

$$n_{glass} \approx 1.5 > n_{air} \approx 1$$

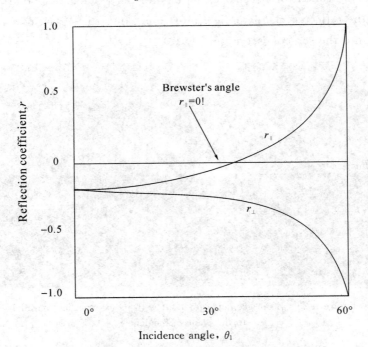

Figure 1.23 Reflection coefficients for an air-to-glass interface

Total internal reflection above the **critical angle**

$$\theta_{crit} \equiv \arcsin\left(\frac{n_t}{n_i}\right)$$

(The sine in Snell's Law can't be > 1!):

$$\sin\theta_{crit} = \frac{n_t}{n_i \sin 90°}$$

Figure 1.24 shows reflection coefficients for a glass-to-air interface.

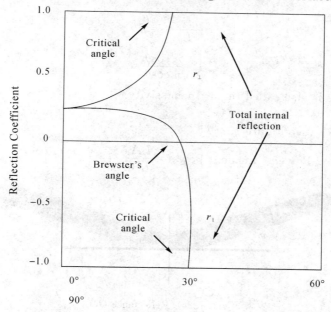

Figure 1.24 Reflection coefficients for a glass-to-air interface

Transmittance (T) (see Figure 1.25)

$T \equiv$ Transmitted Power / Incident Power $= \dfrac{I_t A_t}{I_i A_i}$, $I = \left(n \dfrac{\varepsilon_0 c_0}{2}\right) |E_0|^2$, $A =$ Area

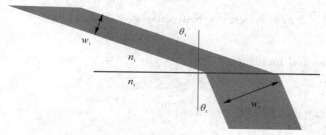

Figure 1.25 Transmittance (T)

Compute the ratio of the beam areas:

$$\frac{A_t}{A_i} = \frac{w_t}{w_i} = \frac{\cos\theta_t}{\cos\theta_i} = m \tag{1.37}$$

Optical Sensing and Measurement

The beam expands in one dimension on refraction.

$$T = \frac{I_t A_t}{I_i A_i} = \frac{\left(n_t \frac{\varepsilon_0 c_0}{2}\right)|E_{0t}|^2}{\left(n_i \frac{\varepsilon_0 c_0}{2}\right)|E_{0i}|^2}\left[\frac{w_t}{w_i}\right] = \frac{n_t}{n_i}\frac{|E_{0t}|^2}{|E_{0i}|^2}\frac{w_t}{w_i} = \frac{n_t}{n_i}t^2\frac{\cos(\theta_t)}{\cos(\theta_i)} \quad (1.38)$$

$$\frac{|E_{0t}|^2}{|E_{0i}|^2} = t^2$$

$$T = \frac{n_t \cos\theta_t}{n_i \cos\theta_i} t^2$$

The Transmittance is also called the Transmissivity.

Reflectance (R) (see Figure 1.26)

$$R \equiv \text{Reflected Power} / \text{Incident Power} = \frac{I_r A_r}{I_i A_i}, \quad I = \left(n\frac{\varepsilon_0 c_0}{2}\right)|E_0|^2, \quad A = \text{Area}$$

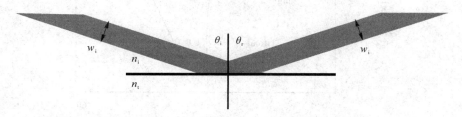

Figure 1.26 Reflectance (R)

Because the angle of incidence = the angle of reflection, the beam area doesn't change on reflection.

Also, n is the same for both incident and reflected beams.

$$\text{So}: R = r^2$$

The Reflectance is also called the Reflectivity.

Reflection at normal incidence

When $\theta_i = 0$,

$$R = \left(\frac{n_t - n_i}{n_t + n_i}\right)^2 \quad (1.39)$$

and

$$T = \frac{4 n_t n_i}{(n_t + n_i)^2} \quad (1.40)$$

For an air-glass interface ($n_i = 1$ and $n_t = 1.5$), $R = 4\%$ and $T = 96\%$

The values are the same, whichever direction the light travels, from air to glass or from glass to air.

The 4% has big implications for photography lenses.

1.7 Fermat's Principle

The law of reflection and the law of refraction tell us how light waves behave at the boundary between two media with different indices of refraction. In 1650, Fermat discovered a way to explain reflection and refraction as the consequence of one single principle. It is called the principle of least time or **Fermat's principle**.

Assume we want light to get from point A to point B, subject to some boundary conditions. For example, we want the light to bounce off a mirror or to pass through a piece of glass on its way from A to B. Fermat's principle states that of all the possible paths the light might take, that satisfy those boundary conditions, **light takes the path which requires the shortest time**.

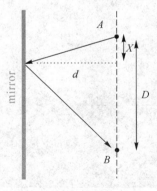

Figure 1.27　Light takes the path which requires the shortest time

Consider the diagram above, Figure 1.27. We want light to leave point A, bounce off the mirror, and get to point B. Let the perpendicular distance from the mirror of both A and B be d and the shortest distance between the points be D. Assume that light takes the path shown. The length of this path is

$$L = (x^2 + d^2)^{1/2} + ((D-x)^2 + d^2)^{1/2} \tag{1.41}$$

Since the speed of light is the same everywhere along all possible paths, the shortest

Optical Sensing and Measurement

path requires the shortest time. To find the shortest path, we differentiate L with respect to x and set the result equal to zero. (This yields an extremum in the function $L(x)$.)

$$\frac{dL}{dx} = \frac{x}{\sqrt{x^2+d^2}} - \frac{D-x}{\sqrt{(D-x)^2+d^2}} = 0 \tag{1.42}$$

$$\frac{x^2}{x^2+d^2} = \frac{D^2+x^2-2Dx}{D^2+x^2-2Dx+d^2} \tag{1.43}$$

$$x^2(D^2+x^2-2Dx+d^2) = (D^2+x^2-2Dx)(x^2+d^2) \tag{1.44}$$

After canceling equal terms on both sides we are left with

$$d^2 D^2 = 2Dx, \quad \text{or} \quad x = D/2 \tag{1.45}$$

The path that takes the shortest time is the one for which $x = D/2$, or equivalently, the one for which $\theta_i = \theta_r$. **Fermat's principle yields the law of reflection.**

Now assume we want light to propagate from point A to point B across the boundary between medium 1 and medium 2.

For the path shown in the figure above (see Figure 1.28) the time required is

$$t = \frac{\sqrt{x^2+d^2}}{c/n_1} - \frac{\sqrt{(D-x)^2+d^2}}{c/n_2} \tag{1.46}$$

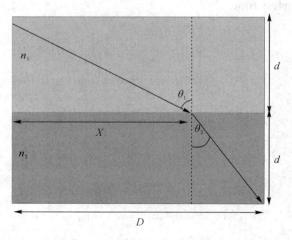

Light propagate across the boundary between medium 1 and medium 2.

Setting $dt/dx = 0$ we obtain

$$n_1 \frac{x}{\sqrt{x^2+d^2}} = n_2 \frac{D-x}{\sqrt{(D-x)^2+d^2}} \tag{1.47}$$

or

$$n_1 \sin\theta_1 = n_2 \sin\theta_2 \tag{1.48}$$

1.7.1 Fermat's principle yields Snell's law

Interference and Diffraction

Light is a transverse electromagnetic wave. The electric field of an electromagnetic plane wave is of the form

$$E(r, t) = E_{max} \cos(k \cdot r - \omega t + \phi), \quad E_{max} \perp k \tag{1.49}$$

The electric field of an EM plane wave traveling in the x-direction therefore may be written as

$$E(x, t) = E_{max} \cos(kx - \omega t + \phi), \quad E_{max} \perp i \tag{1.50}$$

If we consider a linearly polarized plane wave and we orient our coordinate system so that E_{max} is directed along the y-axis then we can drop the vector notation and write

$$E(x, t) = E_{max} \cos(kx - \omega t + \phi), \quad E = E_y \tag{1.51}$$

A more convenient notation is complex notation. For any angle θ,

$$\cos\theta = \text{Re}(\exp(i\theta)), \quad \exp(i\theta) = \cos\theta + i\sin\theta \tag{1.52}$$

We therefore rewrite the above equation as

$$E(x, t) = E_{max} \exp(i(kx - \omega t + \phi)), \quad E = E_y \tag{1.53}$$

When we use complex notation to specify the electric field of an electromagnetic wave, **it is implied that only the real part of the equation describes the electric field.**

Two or more waves traveling in the same medium travel independently and can pass through each other. In regions where they overlap we only observe a single disturbance. We observe interference. When two or more light waves interfere, **the resulting electric field amplitude is equal to the vector sum of the individual field amplitudes.** If two waves with equal amplitudes are **in phase** and overlap i.e. if crest meets crest and trough meets trough, then we observe a resultant wave with twice the amplitude. We have **constructive interference**. If the two overlapping waves, however, are completely **out of phase**, i.e. if crest meets trough, then the two waves cancel each other out completely. We have **destructive interference**.

Let us add two linearly polarized plane electromagnetic waves with equal wavelengths but different phases, traveling along the x-axis.

$$E_1(x, t) = A_1 \exp(i(kx - \omega t + \phi_1)), \quad E_2(x, t) = A_2 \exp(i(kx - \omega t + \phi_2)),$$
$$E(x, t) = E_1(x, t) + E_2(x, t) = (A_1 \exp(i\phi_1) + A_2 \exp(i\phi_2)) \exp(i(kx - \omega t))$$
$$= A_R \exp(i(kx - \omega t + \phi_R)) \tag{1.54}$$

The result of the addition is a linearly polarized plane electromagnetic waves with the same wavelength but a different phase and amplitude, traveling along the x-axis shown in Figure 1.29.

Here A_R is the resultant amplitude and ϕ_R is the resultant phase.

$$A_1 \exp(i\phi_1) + A_2 \exp(i\phi_2) = A_R \exp(i\phi_R) \tag{1.55}$$

To find the magnitude of a complex number we multiply the number by its complex conjugate and then take the square root. The square of the resultant amplitude is given by

$$\begin{aligned}A_R^2 &= (A_1 \exp(i\phi_1) + A_2 \exp(i\phi_2))(A_1 \exp(-i\phi_1) + A_2 \exp(-i\phi_2)) \\ &= A_1^2 + A_2^2 + A_1 A_2 (\exp(i(\phi_1 - \phi_2)) + \exp(-i(\phi_1 - \phi_2))) \\ &= A_1^2 + A_2^2 + 2A_1 A_2 \cos(\phi_1 - \phi_2)\end{aligned} \tag{1.56}$$

The intensity of a wave is proportional to the square of its amplitude. The intensity of the resultant wave is therefore proportional to A_R^2.

To find ϕ_R we use

$$\tan\phi_R = \frac{A_1 \sin\phi_1 + A_2 \sin\phi_2}{A_1 \cos\phi_1 + A_2 \cos\phi_2} \tag{1.57}$$

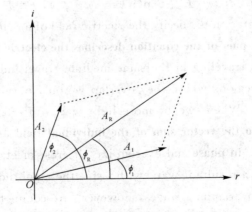

Figure 1.29 Addition of two linearly polarized plane electromagnetic waves

Let us add two linearly polarized plane electromagnetic waves with equal amplitudes but slightly different wavelengths and therefore different frequencies. ($\omega/k = c/n$.)

$$E_1(x, t) = A \exp(i(kx - \omega t)), \quad E_2(x, t) = A \exp(i((k+\delta k)x - (\omega + \delta\omega)t))$$

$$E(x, t) = E_1(x, t) + E_2(x, t) = A \exp(i(kx - \omega t))(1 + \exp(i(\delta k x - \delta\omega t)))$$

$$= A \exp(i(kx - \omega t)) \exp\left(\left(\frac{i}{2}\right)(\delta k x - \delta\omega t)\right) 2\cos\left(\frac{\delta k x - \delta\omega t}{2}\right)$$

$$=2A\ \exp\left(i\left(\left(k+\frac{\delta k}{2}\right)x-\left(\omega+\frac{\delta\omega}{2}t\right)\cos\left(\frac{\delta kx-\delta\omega t}{2}\right)\right)\right) \tag{1.58}$$

We obtain a traveling wave $\exp(i((k+\delta k/2)x-(\omega+Z\omega/2t))$ with the average frequency of the two waves being added with an amplitude $2A\cos((\delta kx-\delta\omega t)/2)$. The amplitude itself is a traveling wave, traveling with speed $v_g = \delta\omega/\delta k$.

We call $v_p = \omega/k = c/n$ the **phase velocity** of the waves and $v_g = \delta\omega/\delta k$ the **group velocity**. For electromagnetic waves in empty space $v_p = v_g = c$. If we have dispersion in a material and the index of refraction $n = n(k)$, the group and phase velocity are not equal to each other.

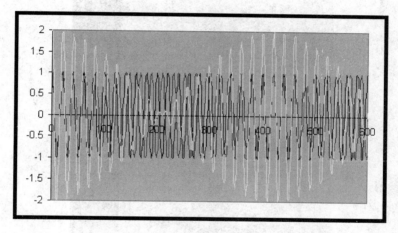

Figure 1.30 The two waves have wavelengths of 20 and 21 units respectively

The two waves in the figure above (see Figure 1.30) have wavelengths of 20 and 21 units respectively. Their frequencies are $v/20$ and $v/21$, where v is the speed of the waves. The frequency of the intensity variations is $v/20 - v/21 = v/420$.

The double slit (interference by division of wavefront)

Interference patterns are only observed if the interfering light from the various sources is **coherent**, i.e. if the phase difference between the sources is constant. Splitting the light from a single source into various beams guaranties coherence as long as the optical path lengths are nearly equal. Light from two different light bulbs is incoherent and will not produce an interference pattern. Lasers are sources of monochromatic, (single wavelength), coherent light. Two lasers can maintain a constant phase difference between each other for relatively long time intervals or for relatively large path-length differences.

Optical Sensing and Measurement

A scheme for thinking about the nature of wave propagation is called **Huygen's principle.** If light from a distant point source is incident onto an obstacle which contains two very small slits a distance d apart, then the wavelets emanating from each slit will constructively interfere behind the obstacle.

If we let the light fall onto a screen behind the obstacle, we will observe a pattern of bright and dark stripes on the screen. This pattern of bright and dark lines is known as a **fringe pattern.** The bright lines indicate constructive interference and the dark lines indicate destructive interference.

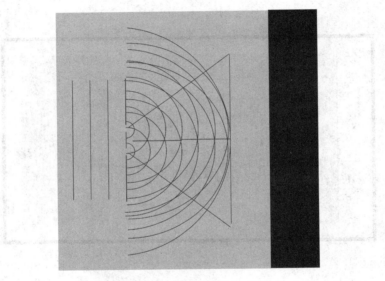

Figure 1.31 Constructive interference of the light from the two slits

The bright fringe in the middle of the diagram above (see Figure 1.31) is caused by constructive interference of the light from the two slits traveling the same distance to the screen. It is known as the **zero-order fringe.** Crest meets crest and trough meets trough. The dark fringes on either side of the zero-order fringe are caused by destructive interference. Light from one slit travels a distance that is 1/2 wavelength longer than the distance traveled by light from the other slit. Crests meets troughs at these locations. The dark fringes are followed by the **first-order fringes,** one on each side of the zero-order fringe. Light from one slit travels a distance that is one wavelength longer than the distance traveled by light from the other slit to reach these positions. Crest again meets crest.

The diagram, Figure 1.32 shows the geometry for the fringe pattern. We assume $L \gg d$. If light with wavelength λ passes through two slits separated by a distance d, we will observe constructively interference at certain angles. These angles are found by applying the condition for constructive interference, which is

$$d\sin\theta = m\lambda, \quad m = 0, 1, 2, \cdots \tag{1.59}$$

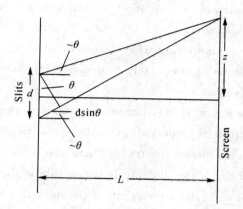

Figure 1.32 The geometry for the fringe pattern

The angles at which dark fringes occur can be found be applying the condition for destructive interference, which is

$$d\sin\theta = \left(m + \frac{1}{2}\right), \quad m = 0, 1, 2, \cdots \tag{1.60}$$

If the interference pattern is viewed on a screen a distance L from the slits, then the wavelength can be found from the spacing of the fringes. We approximately have $\sin\theta = z/L$ and

$$\lambda = \frac{zd}{mL} \tag{1.61}$$

where z is the distance from the center of the interference pattern to the mth bright line in the pattern. This applies as long as the angle θ is small, i.e., as long as z is small compared to L.

What happens when light encounters an entire array of identical, equally-spaced slits, called a diffraction grating?

The bright fringes, which come from constructive interference of the light waves from different slits, are found at the same angles they are found if there are only two slits. But the pattern is much sharper. Why?

For two slits, there is one single position between bright peaks, where the interference is totally destructive. Between the zero-order and first-order fringes, there is one position which requires that one of the waves travels exactly 1/2 wavelength further than the other to reach it. For three slits, however, there are two positions where destructive interference takes place. One is located at the point where the path lengths differ by 1/3 of a wavelength, while the other is located where the path lengths differ by 2/3 of a wavelength. For 4 slits, there are three positions, for 5 slits there are four positions, etc. For a diffraction grating with a large number of slits, the pattern is sharp because of the many positions of completely destructive interference between the bright, constructive-interference fringes.

Diffraction gratings(see Figure 1.33), like prisms, disperse white light into its component colors. The spectral pattern is repeated on either side of the main pattern. These repetitions are called "higher order spectra". There are often many of them, each one fainter than the previous one. If the distance between slits is d, and the angle to a bright fringe of a particular color is θ, the wavelength of the light can be calculated.

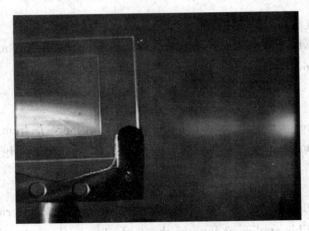

Figure 1.33 Diffraction grating

Thin-film interference (interference by division of amplitude)

When a light wave reflects from a medium with a larger index of refraction, then the phase shift of the reflected wave with respect to the incident wave isp 180°. When a light wave reflects from a medium with a smaller index of refraction, then the phase shift of the reflected wave with respect to the incident wave is zero. When a light wave is reflected, the reflected and the incident wave interfere.

Chapter 1 Nature of Light

Constructive and destructive interference of reflected light waves causes the colorful patterns we often observe in thin films, such as soap bubbles and layers of oil on water. **Thin-film interference** (see Figure 1.34) is the interference of light waves reflecting off the top surface of a film with the waves reflecting from the bottom surface. If the thickness of the film is on the order of the wavelength of light, then colorful patterns can be obtained, as shown in the image below, Figure 1.35.

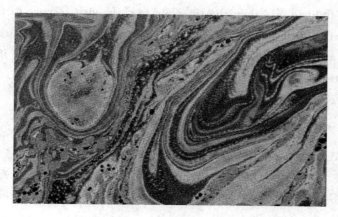

Figure 1.34 Thin-film interference

Consider the case of a thin film of oil of thickness t floating on water. For simplicity, assume that the light is incident normally, so that the angle of incidence and the angle of reflection are zero.

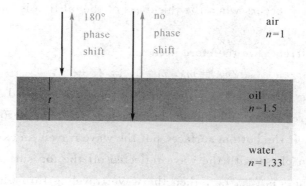

Figure 1.35 The interference of light waves reflecting off the top surface of a film with the waves reflecting from the bottom surface

In the air, the light reflecting off the air-oil interface will have a 180° phase shift with respect to the incident light. A 180° phase shift is equivalent to the light having traveled a distance of 1/2 wavelength. In the oil, the light reflecting from the oil-water interface will have no phase shift with respect to the light incident on the interface. For the light reflected off the oil and the light reflected off the water to constructively interfere we need the two reflected waves to have a phase shift of an integer multiple of 2π (360°). If the light reflected off the oil-water interface travels an additional distance equal to 1/2 the wavelength of the light in oil, then the total phase shift with respect to the light reflected off the air-oil interface will be 2π. This happens if the thickness of the film is equal to 1/4 the wavelength of the light in oil. We also get constructive interference if the thickness of the film is equal to 3/4, 5/4, etc, the wavelength of the light in oil. For constructive interference we need

$$2t = \left(m + \frac{1}{2}\right)\lambda_n, \quad m = 0, 1, 2, \cdots \tag{1.62}$$

where λ_n is the wavelength of the light in oil.

In vacuum we have $\lambda f = c$. In a medium with index of refraction n we have $\lambda_n f = c/n$. The frequency of oscillation is the same in vacuum and in a medium, therefore

$$\lambda_n = \frac{\lambda}{n} \tag{1.63}$$

For constructive interference we therefore need

$$2n_{oil}t = (m + \frac{1}{2})\lambda, \quad m = 0, 1, 2, \cdots \tag{1.64}$$

Destructive interference occurs when the thickness of the oil film is equal to $(1/2)\lambda_n$, λ_n, $(3/2)\lambda_n$, etc.

For destructive interference we therefore need

$$2n_{oil}t = m\lambda, \quad m = 0, 1, 2, \cdots \tag{1.65}$$

If the thickness of the film is $(1/4)\lambda_n$, the phase of the wave reflected off the top surface is shifted by π by the reflection. The phase of the wave traveling through the film is not shifted by reflection off the bottom surface, but the wave travels an extra distance of $\lambda_n/2$. It will therefore be in phase with the wave reflected off the top surface. If, on the other hand, the film thickness is $(1/2)\lambda_n$, then the wave traveling through the film travels an extra distance of 1 wavelength. It will therefore be out of phase with the wave reflected off the top surface and the two waves will cancel each other out.

Waves incident at an angle θ_i on the air oil interface are refracted as they enter the oil.

The angle of refraction θ_t is found from Snell's law, $n_{air}\sin\theta_i = n_{oil}\sin\theta_t$. If they are reflected off the second interface, then they travel a distance $2t/\cos\theta_t$ in the oil. When they emerge again from the oil into the air and propagate parallel to the waves reflected at the air-oil interface, then the total optical path length difference is

$$\frac{2n_{oil}t}{\cos\theta_t} - 2t\,\tan\theta_t\sin\theta_i = \frac{2n_{oil}t}{\cos\theta_t} - 2t\,\tan\theta_t\left(\frac{n_{oil}}{n_{air}}\right)\sin\theta_t \quad (1.66)$$

$$= \frac{2n_{oil}t(1-\sin^2\theta_t)}{\cos\theta_t} = 2n_{oil}t\,\cos\theta_t$$

For constructive interference we therefore need

$$2n_{oil}t\cos\theta_t = (m+\frac{1}{2})\lambda,\ m=0,\ 1,\ 2,\ \cdots \quad (1.67)$$

and for destructive interference we need

$$2n_{oil}t\cos\theta_t = m\lambda,\ m=0,\ 1,\ 2,\ \cdots \quad (1.68)$$

Constructive and destructive interference occur at different angles for different wavelength. The observer sees colored bands(see Figure 1.36).

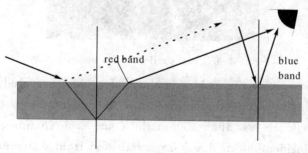

Figure 1.36 Colored bands

The destructive interference of reflected light waves is utilized to make **non-reflective coatings**. Such coatings are commonly found on camera lenses and binocular lenses, and often have a bluish tint. The coating is put over glass, and the coating material generally has an index of refraction less than that of glass. Then the phase shift of both reflected waves is 180°, and a film thickness equal to 1/4 of the wavelength of light in the film produces a net shift of 1/2 wavelength, resulting in cancellation. For such non-reflective coatings the minimum film thickness t required is

$$t = \frac{\lambda}{4n} \quad (1.69)$$

where n is the index of refraction of the coating material. A coating with thickness $t=\lambda/4n$

prevents the reflection of most of the light with a wavelength λ close to $\lambda = 4nt$. The coating does not reflect a specific range of wavelengths. Often that range is chosen to be in the yellow-green region of the spectrum, where the eye is most sensitive. Lenses coated for the yellow-green region reflect in the blue and red regions, giving the surface a familiar purple color.

The colorful patterns that you see when light reflects off a compact disk are produced by thin film interference(see Figure 1.37).

Figure 1.37 Light reflects off a compact disk

A compact disc is made of a polycarbonate wafer which is coated with a metallic film, usually an aluminum alloy. The aluminum film is then covered by a plastic polycarbonate coating. The coatings are less than 100nm thick and each coating partially reflects and partially transmits incident light. Light rays reflected from different coating boundaries interfere with each other to produce the colorful patterns. The reflectance of the CD is not uniform, because CD disk contains a long string of pits written helically on the disk. These pits encode the information stored on the CD(see Figure 1.38).

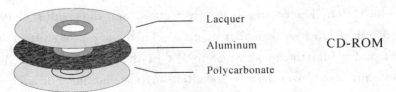

Figure 1.38 The coatings of a compact disc

Thin film interference is used in industry as a non-contact, non-destructive way to measure film thicknesses.

Chapter 1 Nature of Light

Diffraction is the tendency of a wave emitted from a finite source or passing through a finite aperture to spread out as it propagates. Diffraction results from the interference of an infinite number of waves emitted by a continuous distribution of source points. According to Huygens principle every point on a wave front of light can be considered to be a secondary source of spherical wavelets.

The wave fronts of a spherical wave are spherical surfaces surrounding a point source. The wave vector k points away from the point source along the radius of the sphere at any point. If the point source is located at the origin, then k is always parallel to the position vector r, $k \parallel r$, and $k \cdot r = kr$. Here $k = 2\pi/\lambda$ is the wave number and r is just the scalar distance along any radial direction.

The area of a wave front with radius r is $4\pi r^2$. Since energy is conserved, the intensity must decrease with r as $1/r^2$. Since the intensity is proportional to the square of the amplitude, the amplitude must decrease with r as $1/r$, and we have $E(r) = (A/r)\cos(kr - \omega t)$, with A being a constant.

Maxwell's equation can be used to formulate an exact description of the propagation of a light wave through an optical system and the space around it. This is rarely done. Two standard approximations to the exact formulation of wave propagation and diffraction are the Fraunhofer and Fresnel approximations. These approximations ignore the vector nature of the electric field.

The **Fraunhofer regime** is the far-field regime (see Figure 1.39). Very far from a point source the wave fronts are essentially plane waves. The Fraunhofer approximation is only valid when the source, aperture, and detector are all very far apart or when lenses are used to convert spherical waves into plane waves.

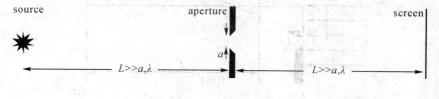

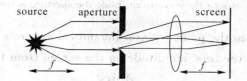

Figure 1.39 The Fraunhofer regime

Optical Sensing and Measurement

The **Fresnel regime** is the near-field regime(see Figure 1.40). In this regime the wave fronts are curved, and their mathematical description is more involved.

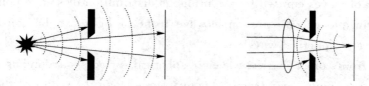

Figure 1.40 Fresnel regime

1.7.2 The single slit

Assume light from a distant source passes through a narrow slit. What do we observe on a distant screen?

According to the Huygen-Fresnel principle, the total field at a point y on the screen is the superposition of wave fields from an infinite number of point sources in the aperture region(see Figure 1.41). Each points on the wave front inside the aperture ($-a/2 \leqslant s \leqslant a/2$) is the source of a spherical wave. A distance r from the point s the electric field is due to this point sources is

$$dE = \left(\frac{A_s ds}{r}\right) \cos(kr - \omega t) \tag{1.70}$$

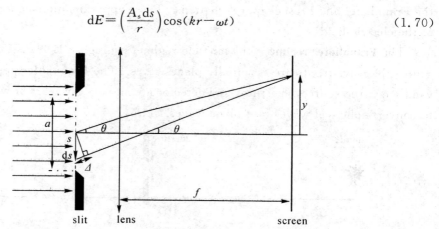

Figure 1.41 Each points on the wave front inside the aperture is the source of a spherical wave

If r_0 is the distance from the point $s=0$ on the optical axis to a point y on the screen, then the contribution dE to the total amplitude on the screen from the point at $s=0$ is

$$dE(y) = \left(\frac{A_s ds}{r_0}\right) \cos(kr_0 - \omega t) \tag{1.71}$$

Here A_s/r_0 is the amplitude per unit width and ds is the infinitely small width of a point source. For off-axis points for which $s\neq 0$, the distance is longer or shorter than r_0 by an amount Δ.

The contribution $dE(y)$ to the total amplitude on the screen from an off-axis point ($s\neq 0$) is

$$dE(y) = \left(\frac{A_s ds}{r_0 + \Delta(s)}\right)\cos(k(r_0 + \Delta(s)) - \omega t) \tag{1.72}$$

To find the total amplitude $E(y)$ we have to add up the contributions from all points on the aperture. Because there are an infinite number of points, the sum becomes an integral.

$$E(y) = \int_{-\frac{\partial}{2}}^{\frac{\partial}{2}} \frac{A_s}{r_0 + \Delta(s)}\cos[k(r_0 + \Delta(s)) - \omega t] ds \tag{1.73}$$

We define $\sin\theta = \Delta/s$. Since $r_0 \gg \Delta$, we approximate $1/(r_0 + \Delta)$ with $1/r_0$. However we cannot drop the Δ inside the cosine function, since $k\Delta(s)$ is not necessarily much smaller than 2π.

We then have

$$E(y) = \frac{A_s}{r_0}\int_{-\frac{\partial}{2}}^{\frac{\partial}{2}} \cos[(k\sin\theta)s + (kr_0 - \omega t)] ds \tag{1.74}$$

Using

$$\int \cos(ax + b) = \frac{1}{a}\sin(ax + b) \tag{1.75}$$

integration then yields

$$E(y) = \frac{\sin(k\partial \sin\theta/2)}{k\partial \sin\theta/2}\frac{A_s \partial}{r_0}\cos(kr_0 - \omega t) \tag{1.76}$$

The function $\sin x/x = \text{sinc } x$ is called the sinc function (see Figure 1.42).

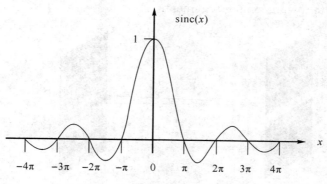

Figure 1.42 The sinc function

The intensity is proportional to the square of the field,

$$I(y) \propto E^2(y) = \frac{\sin^2(k\partial \sin\theta/2)}{(k\partial \sin\theta/2)^2}\left(\frac{A_s \partial}{r_0}\right)^2 \cos^2(kr_0 - \omega t) \tag{1.77}$$

Since the square of a cosine function averages to $^{1/2}$, the time-averaged intensity(see Figure 1.43) is given by

$$\langle I(y)\rangle = \frac{\sin^2(\pi\,\partial\sin(\theta)/\lambda)}{(\pi\,\partial\sin(\theta)/\lambda)^2}\langle I(0)\rangle \tag{1.78}$$

where $\langle I_0\rangle \propto (1/2)(A_s a/r_0)^2$ and $k = 2\pi/\lambda$.

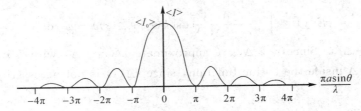

Figure 1.43 The time-averaged intensity

The time-averaged intensity has a peak in the center with smaller fringes on the sides.

For small angles we may approximate $\sin\theta \cong \theta$. the sinc function Then the first zeros on the sides of the central peak occur when

$$\frac{a\sin\theta}{\lambda} \cong \frac{\pi a\theta}{\lambda} = \pi, \quad \text{or} \quad \theta = \frac{\lambda}{a}$$

On the screen we see a pattern similar to that shown in the figure(see Figure 1.44) below. The positions of all maxima and minima in the diffraction pattern from a single slit can also be found from the following simple arguments.

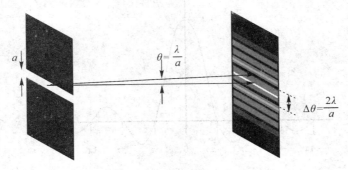

Figure 1.44 A pattern on the screen

Chapter 1 Nature of Light

When light passes through a single slit whose width w is on the order of the wavelength of the light, then we observe a single slit diffraction pattern. Huygen's principle tells us that each part of the slit can be thought of as an emitter of waves. All these waves interfere to produce the diffraction pattern. Consider a slit of width w as shown in the diagram below(See Figure 1.45).

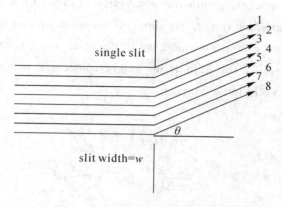

Figure 1.45 Light passes through a single slit

For light leaving the slit in a particular direction, we may have destructive interference between the ray at the top edge, consider a wave propagating through space. Coherence is a measure of the correlation that exists between the phases of the wave measured at different points. The coherence of a wave depends on the characteristics of its source.

Let us look at a simple example. Imagine two corks bobbing up and down on a wavy water surface. Suppose the source of the water waves is a single stick moved harmonically in and out of the water, breaking the otherwise smooth water surface. There exists a perfect correlation between the motions of the two corks. They may not bop up and down exactly in phase, one may go up while the other one goes down, but the phase difference between the positions of the two corks is constant in time. We say that the source is perfectly coherent. A harmonically oscillating point source produces a perfectly coherent wave.

When we describe the coherence of light waves, we distinguish two types of coherence.

Temporal coherence is a measure of the correlation between the phases of a light wave at different points along the direction of propagation. Temporal coherence tells us how monochromatic a source is.

Assume our source emits waves with wavelength $\lambda \pm \Delta\lambda$. Waves with wavelength λ and $\lambda + \Delta\lambda$, which at some point in space constructively interfere, will destructively interfere after some optical path length $l_c = \lambda^2/(2\pi\Delta\lambda)$; l_c is called the coherence length. The phase of a wave propagating into the x-direction is given by $\phi = kx - \omega t$. Look at the wave pattern in space at some time t. At some distance l the phase difference between two waves with wave vectors k_1 and k_2 which are in phase at $x=0$ becomes $\Delta\phi = (k_1 - k_2)l$. When $\Delta\phi = 1$, or $\Delta\phi \approx 60°$, the light is no longer considered coherent. Interference and diffraction patterns severely loose contrast.

We therefore have

$$l = (k_1 - k_2)l_c = \left(\frac{2\pi/\lambda - 2\pi}{\lambda + \Delta\lambda}\right)l_c$$

$$\frac{(\lambda + \Delta\lambda - \lambda)l_c}{\lambda(\lambda + \Delta\lambda)} \approx \frac{\Delta\lambda l_c}{\lambda^2} = \frac{1}{2}\pi$$

$$l_c = \frac{\lambda^2}{2\pi\Delta\lambda}$$

The wave pattern travels through space with speed c.
The coherence time t_c is $t_c = l_c/c$. Since $\lambda f = c$, we have $\Delta f/f = \Delta\omega/\omega = \Delta\lambda/\lambda$. We can write

$$l_c = \frac{\lambda^2}{2\pi\Delta\lambda} = \frac{\lambda f}{2\pi\Delta f} = \frac{c}{\Delta\omega}$$

$$t_c = \frac{1}{\Delta\omega}$$

If we know the wavelength or frequency spread of a light source, we can calculate l_c and t_c. We cannot observe interference patterns produced by division of amplitude, such as thin-film interference if the optical path difference greatly exceeds l_c.

Spatial coherence is a measure of the correlation between the phases of a light wave at different points transverse to the direction of propagation. Spatial coherence tells us how uniform the phase of the wave front is.

A distance L from a source whose linear dimensions are on the order of δ, two slits separated by a distance greater than $d_c = 0.16\lambda L/\delta$ will no longer produce an interference pattern. We call $\pi d_c^2/4$ the coherence area of the source (see Figure 1.46).

At time t look at a source of width δ a perpendicular distance L from a screen. Look at

Chapter 1   Nature of Light

two points (P1 and P2) on the screen separated by a distance d.

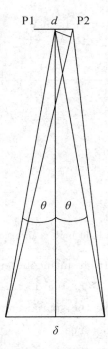

Figure 1.46   The coherence area of the source

Light waves emitted from the two edges of the source have a some definite phase difference right in the center between the to points at some time t. A ray traveling from the left edge of δ to point P2 must travel a distance $d(\sin\theta)/2$ farther then a ray traveling to the center. The path of a ray traveling from the right edge of δ to point P2 travel is $d(\sin\theta)/2$ shorter then the path to the center. The path difference for the two rays therefore is $d\sin\theta$, which introduces a phase difference $\Delta\phi' = 2\pi d\sin\theta/\lambda$. For the distance from P1 to P2 we therefore get a phase difference $\Delta\phi = 2\Delta\phi' = 4\pi d\sin\theta/\lambda$. Wavelets emitted from the two edges of the source are that are in phase at P1 at time t are are out of phase by $4\pi\, d\sin\theta/\lambda$ at P2 at the same time t. We have $\sin\theta \approx \delta/(2L)$, so $\Delta\phi = 2\pi d\delta/(L\lambda)$. When $\Delta\phi = 1$ or $\Delta\phi \approx 60°$, the light is no longer considered coherent.

$$\Delta\phi = 1 \rightarrow d = \frac{L\lambda}{2\pi\delta} = \frac{0.16\, L\lambda}{\delta}$$

An incandescent light bulb is an example of very incoherent source(see Figure 1.47).

Optical Sensing and Measurement

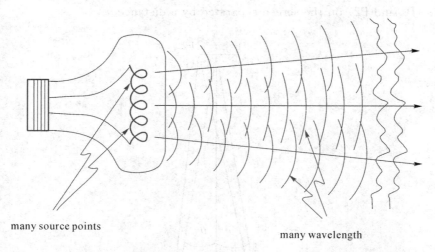

many source points   many wavelength

Figure 1.47 An incandescent light bulb is an example of very incoherent source

We can produce coherent light from an incoherent source if we are willing to throw away a lot of the light (see Figure 1.48). We do this by first spatially filtering the light from the incoherent source to increase the spatial coherence, and then spectrally filtering the light to increase the temporal coherence.

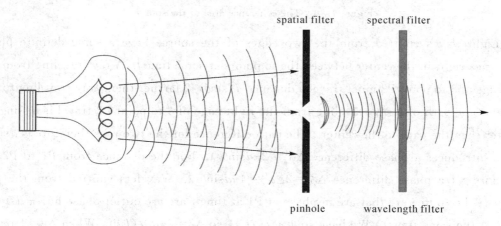

Figure 1.48 Produce coherent light from an incoherent source

A sinusoidal plane wave extends to infinity in space and time. It is perfectly coherent in space and time, its coherence length, coherence time, and coherence area are all infinite. All real waves are wave pulses, they last for a finite time interval and have finite extend perpendicular to their direction of propagation. They are mathematically described by non-periodic functions. We therefore have to learn how to analyze non-periodic functions to find the frequencies present in wave pulses to determine $\Delta\omega$ and the coherence length.

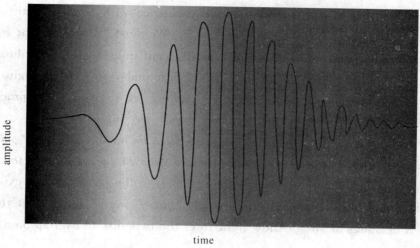

Figure 1.49   The light pulse

The light pulse in the figure above contains many frequencies (see Figure 1.49). To determine the coherence length, we need to know its frequency contend.

Any set of sinusoidal waves whose frequencies belong to a **harmonic series** will combine to produce a **periodic complex wave**, whose repetition frequency is that of the series fundamental. The individual components may have any amplitudes and any relative phases. These amplitudes and phases determine the shape of the complex waveform.

1.8   Interference

Today we produce interference effects with little difficulty. In the days of Sir Isaac Newton and Christian Huygens, however, light interference was not easily demonstrated. There were several reasons for this. One was based on the extremely short wavelength of visible light, around 20 millionths of an inch, and the obvious difficulty associated with seeing or detecting interference patterns formed by overlapping waves of so short a wavelength, and so rapid a vibration, around a million billion cycles per second! Another reason was based on the difficulty, before the laser came along, of creating *coherent* waves, that is, waves with a phase relationship with each other that remained *fixed* during the time when interference was observed.

It turns out that we *can* develop phase coherence with *nonlaser* light sources to demonstrate interference, but we must work at it. We must "prepare" light from readily available incoherent light sources, which typically emit individual, uncoordinated, short wave trains of fixed phase of no longer than 10^{-8} seconds, so that the light from such sources remains coherent over periods of time long enough to overlap and produce visible interference patterns. There are generally two ways to do this.

- Develop several coherent *virtual* sources from a single incoherent "point" source with the help of mirrors. Allow light from the two virtual sources to overlap and interfere. (This method is used, for example, in the Loyd's mirror experiment.)
- Take monochromatic light from a single "point" source and pass it through two small openings or slits. Allow light from the two slits to overlap on a screen and interfere.

We shall use the second of these two methods to demonstrate Thomas Young's famous *double-slit experiment*, worked out for the first time at the very beginning of the 19th century. But first, let's consider the basics of interference from two point sources.

1.8.1 Constructive and destructive interference

Figure 1.50 shows two "point" sources of light, S and S', whose radiating waves maintain a fixed phase relationship with each other as they travel outward. The emerging waves are in effect spherical, but we show them as circular in the two-dimensional drawing. The solid circles represent crests, the dashed circles, troughs.

In Figure 1.50, along directions OP, OP'_2, and OP_2 (emphasized by solid dots) crests from S and S' meet (as do the troughs), thereby creating a condition of *constructive interference*. As a result, light striking the screen at points P, P_2, and P'_2 is at a maximum intensity and a bright spot appears. By contrast, along directions OP_1 and OP'_1 (emphasized by open circles) crests and troughs meet each other, creating a condition of *destructive interference*. So at points P_1 and P'_1 on the screen, no light appears, leaving a dark spot.

The requirement of *coherent sources is* a stringent requirement if interference is to be observed. To see this clearly, suppose for a moment that sources S and S' in Figure 1.50 are, in fact, *two corks* bobbing up and down on a quiet pond. As long as the two corks maintain a *fixed* relationship between their vertical motions, each will produce a series of *related* crests and troughs, and observable interference patterns in the overlapregion will occur. But if the two corks bob up and down in a *random, disorganized manner*, no series of related, fixed-phase crests and troughs will form and no interference patterns of sufficiently long duration can develop, and so interference will not be observed.

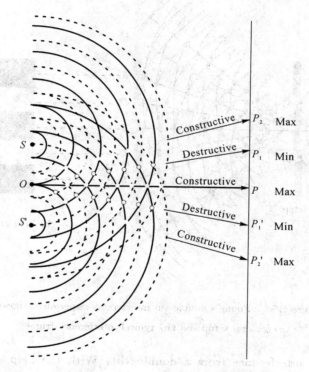

Figure 1.50 Wave interference created by overlapping waves from coherent sources S and S'

1.8.2 Young's double-slit interference experiment

Figure 1.51 shows the general setup for producing *interference* with coherent light from two slits S_1 and S_2. The source S_0 is a monochromatic point source of light whose spherical wave fronts (circular in the drawing) fall on the two slits to create secondary sources S_1 and S_2. Spherical waves radiating out from the two secondary sources S_1 and S_2 maintain a fixed phase relationship with each other as they spread out and overlap on the screen, to produce a series of alternate bright and dark regions, as we saw in Figure 1.50. The alternate regions of bright and dark are referred to as *interference fringes*. Figure 1.51 shows such interference fringes, greatly expanded, for a small central portion of the screen shown in Figure 1.51.

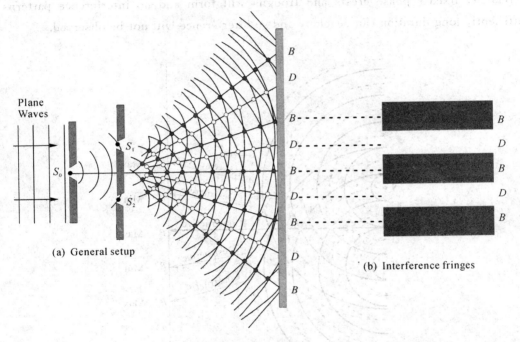

Figure 1.51 Young's double-slit interference experiment showing
(a) general setup and (b) typical interference fringes

Detailed analysis of interference from a double slit: With the help of the *principle of superposition*, we can calculate the positions of the alternate maxima (bright regions) and minima (dark regions) shown in Figure 1.51. To do this we shall make use of Figure 1.52 and the following conditions:

(a) Light from slits S_1 and S_2 is coherent; that is, there exists a fixed phase relationship between the waves from the two sources.

(b) Light from slits S_1 and S_2 is of the same wavelength.

Chapter 1 Nature of Light

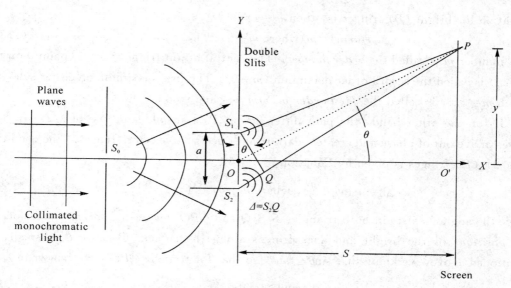

Figure 1.52 Schematic for double-slit interference calculations. Source S_0 is generally a small hole or narrow slit; sources S_1 and S_2 are generally long, narrow slits perpendicular to the page.

In Figure 1.52, light waves from S_1 and S_2 spread out and overlap at an arbitrary point P on the screen. If the overlapping waves are in phase, we expect a bright spot at P; if they are out of phase, we expect a dark spot. So the *phase difference* between the two waves arriving at point P is a key factor in determining what happens there. We shall express the phase difference in terms of the *path difference*, which we can relate to the wavelength λ.

For clarity, Figure1.52 is not drawn to scale. It will be helpful in viewing the drawing to know that, in practice, the distance s from the slits to the screen is about *one meter*, the distance a between slits is *less than a millimeter*, so that the angle θ in triangle $S_1 S_2 Q$, or triangle OPO', is quite small. And on top of all this, the wavelength of light is a fraction of a micrometer.

The path difference Δ between $S_1 P$ and $S_2 P$, as seen in Figure 1.52, is given by Equation 1-80, since the distances PS_1 and PQ are equal and since $\sin\theta = \Delta/a$ in triangle $S_1 S_2 Q$.

$$\Delta = S_2 P - S_1 P = S_2 Q = a \sin\theta \qquad (1.79)$$

If the path difference Δ is equal to λ or some integral multiple of λ, the two waves arrive at P in phase and a bright fringe appears there (constructive interference). The condition for bright (B) fringes is, then,

$$\Delta_B = a\sin\theta = m\lambda \text{ where } m = 0, \pm 1, \pm 2, \ldots \qquad (1.80)$$

The number m is called the *order number*. The central bright fringe at $\theta=0$ (point 0 on the screen) is called the zeroth-order maximum ($m=0$). The first maximum on either side, for which $m=\pm 1$, is called the *first-order maximum*, and so on.

If, on the other hand, the path difference at P is an odd multiple of $\lambda/2$, the two waves arrive out of phase and create a dark fringe (destructive interference). The condition for dark (D) fringes is given by Equation.

$$\Delta_D = a\sin\theta = \frac{m\lambda}{2}, \text{ where } m = 0, \pm 1, \pm 3, \ldots \qquad (1.81)$$

Since the angle θ exists in both triangles $S_1 S_2 Q$ and OPO', we can find an expression for the *positions* of the bright and dark fringes along the screen. Because θ is small, as mentioned above, we know that $\sin\theta \cong \tan\theta$, so that for triangle OPO', we can write

$$\sin\theta \cong \tan\theta = \frac{y}{\lambda s} \qquad (1.82)$$

Combining Equation (1.82) with Equations (1.81) and (1.80) in turn, by substituting for $\sin\theta$ in each, we obtain expressions for the position y of bright and dark fringes on the screen.

$$y_B = \frac{\lambda s}{a} m, \text{ where } m = 0, \pm 1, \pm 2 \qquad (1.83)$$

$$y_D = \frac{\lambda s}{a}\left(m + \frac{1}{2}\right), \text{ where } m = 0, \pm 1, \pm 2, \qquad (1.84)$$

In Example 2, through the use of Equation (1.82), we recreate the method used by Thomas Young to make the *first* measurement of the wavelength of light.

1.9 Diffraction

1.9.1 Diffraction

The ability of light to bend around corners, a consequence of the wave nature of light, is fundamental to both interference and diffraction. *Diffraction* is simply any deviation from geometrical optics resulting from the *obstruction* of a wave front of light by some obstacle or some opening. Diffraction occurs when light waves pass through small openings, around obstacles, or by sharp edges.

Several common diffraction patterns as sketched by an artist are shown in Figure 1.53. Figure 1.53(a) is a typical diffraction pattern for HeNe laser light passing through a circular pinhole. Figure 1.53(b) is a typical diffraction pattern for HeNe laser light passing through a narrow (vertical) slit. And Figure 1.53(c) is a typical pattern for diffraction by a sharp edge.

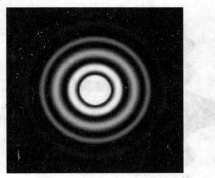
(a) Pinhole diffraction

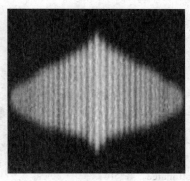
(b) Single-slit diffraction

(c) Straight-edge diffraction

Figure 1.53 Sketches of several common diffraction patterns

The intricacy of the patterns should convince us, once and for all, that geometrical ray optics is incapable of dealing with diffraction phenomena. To demonstrate how wave theory does account for such patterns, we now examine the phenomenon of diffraction of waves by a single slit.

1.9.2 Diffraction by a single slit

The overall geometry for diffraction by a single slit is shown in Figure 1.54. The slit opening, seen in cross section, is in fact a long, narrow slit, perpendicular to the page. The shaded "humps" shown along the screen give a rough idea of intensity variation in the pattern, and the sketch of bright and dark regions to the right of the screen simulates the actual fringe pattern seen on the screen. We observe a wide central bright fringe, bordered by narrower regions of dark and bright. The angle θ shown connects a point P on the screen to the center of the slit.

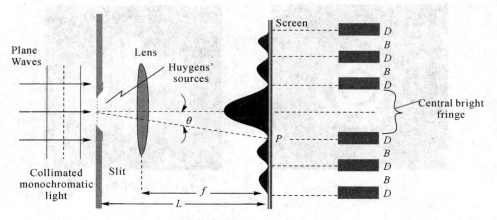

Figure 1.54 Diffraction pattern from a single slit

Since plane waves are incident on the screen, the diffraction pattern in the absence of the focusing lens would be formed far away from the slit and be much more spread out than that shown in Figure 1.54. The lens serves to focus the light passing through the slit onto the screen, just a focal length f away from the lens, while preserving faithfully the relative details of the diffraction pattern that would be formed on a distant screen without the lens.

To determine the location of the minima and maxima on the screen, we divide the slit opening through which a plane wave is passing into many point sources (Huygens' sources), as shown by the series of tiny dots in the slit opening of Figure1.54. These numerous point sources send out Huygens' spherical waves, all in phase, toward the screen. There, at a point such as P, light waves from the various Huygens' sources overlap and interfere, forming the variation in light intensity shown in Figure 1.54. Thus, diffraction considers the contribution from *every part of the wave front* passing through the aperture. By contrast, when we looked at interference from Young's double slit, we considered *each slit* as a point source, ignoring details of the portions of the wave fronts in the slit openings themselves.

The mathematical details involved in adding the contributions at point P from each of the Huygens' sources can be found in basic texts on *physical optics*. Here we give only the end result of the calculation. Equation 1.85 locates the minima, y_{min}, on the screen, in terms of the slit width b, slit-to-screen distance L, wavelength λ, and order m.

$$y_{min} = \frac{m\lambda L}{b}, \text{ where } m = 1, 2, 3, \ldots \qquad (1.85)$$

Figure 1.55 shows the positions of several orders of minima and the essential parameters associated with the single-slit diffraction pattern. (The positions of the *maxima* are mathematically more complicated to express, so we typically work with the positions of the well-defined minima.)

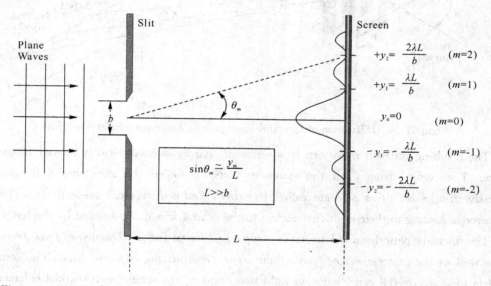

Figure 1.55 Positions of adjacent minima in the diffraction patterns (Drawing is not to scale.)

Now let's use Equation (1.85) to work several sample problems.

1.9.3 Diffraction grating

If we prepare an aperture with thousands of adjacent slits, we have a so-called *transmission-diffraction grating*. The width of a single slit, the opening is given by d, and the distance between slit centers is given by l (see Figure 1.56). For clarity, only a few of the thousands of slits normally present in a grating are shown. Note that the spreading of light occurs always in a direction perpendicular to the direction of the long edge of the slit opening, that is, since the long edge of the slit opening is *vertical* in Figure 1.56, the spreading is in the *horizontal* direction along the screen.

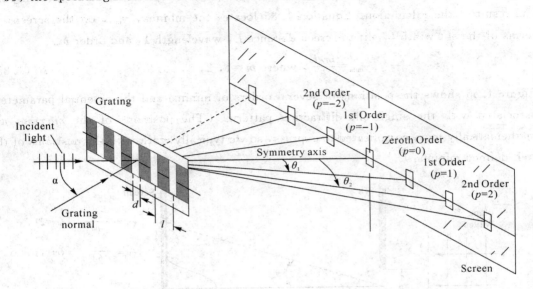

Figure 1.56 Diffraction of light through a grating under Fraunhofer conditions

The resulting diffraction pattern is a series of sharply defined, widely spaced fringes, as shown. The central fringe, on the symmetry axis, is called the *zeroth-order* fringe. The successive fringes on either side are called 1*st order*, 2*nd order*, etc., respectively. They are numbered according to their positions relative to the central fringe, as denoted by the letter *p*.

The intensity pattern on the screen is a *superposition* of the *diffraction effects from each slit* as well as the *interference effects of the light from all the adjacent slits*. The combined effect is to cause overall cancellation of light over most of the screen with marked enhancement over only limited regions, as shown in Figure 1.56. The location of the bright fringes is given by the following expression, called the *grating equation*, assuming that Fraunhofer conditions hold.

$$l(\sin\alpha + \sin\theta_p) = p\lambda, \text{ where } p = 0, \pm 1, \pm 2, \tag{1.86}$$

where l = distance between slit centers

α = angle of incidence of light measured with respect to the normal to the grating surface

θ_p = angle locating the pth-order fringe

p = an integer taking on values of $0, \pm 1, \pm 2,$ etc.

λ = wavelength of light

Note that, if the light is incident on the grating along the grating normal ($\alpha=0$), the grating equation, Equation (1.86), reduces to the more common form shown in Equation (1.87).

$$l(\sin\theta_p) = p\lambda \tag{1.87}$$

If, for example, you shine a HeNe laser beam perpendicularly onto the surface of a transmission grating, you will see a series of brilliant red dots, spread out as shown in Figure 1.56. A complete calculation would show that less light falls on each successively distant red dot or fringe, the $p=0$ or central fringe being always the brightest. Nevertheless, the location of each bright spot, or fringe, is given accurately by Equation (1.86) for either normal incidence ($\alpha=0$) or oblique incidence ($\alpha>0$). If light containing a mixture of wavelengths (white light, for example) is directed onto the transmission grating, Equation (1.86) holds for *each* component color or wavelength. So each color will be spread out on the screen according to Equation (1.86), with the longer wavelengths (red) spreading out farther than the shorter wavelengths (blue). In any case, the central fringe ($p=0$) always remains the same color as the incident beam, since all wavelengths in the $p=0$ fringe have $\theta_p=0$, hence all overlap to re-form the "original" beam and therefore the original "color." Example shows calculations for a typical diffraction grating under Fraunhofer conditions.

According to Fourier analysis, an arbitrary periodical waveform can be regarded as a superposition of sinusoidal waves(see Figure 1.57). **Fourier synthesis** means superimposing many sinusoidal waves to obtain the arbitrary periodic waveform.

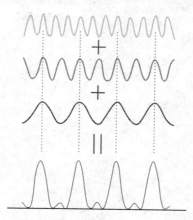

Figure 1.57 Arbitrary periodical waveform can be regarded as a superposition of sinusoidal waves

Optical Sensing and Measurement

Any set of sinusoidal waves whose frequencies do **not** belong to a harmonic series will combine to produce a complex wave that is not periodic. Any non-periodic waveform may be built from a set of sinusoidal waves. Each component must have just the right amplitude and relative phase to produce the desired waveform.

Chapter 2 Polarization

We continue our discussion of the main concepts in *physical optics* with a brief look at *polarization*. Before we describe the polarization of light waves, let's take a look at a simplistic but helpful analogy of "polarization" with *rope waves*.

2.1 Polarization of Light Waves

Imagine a "magic" rope that you can whip up and down at one end, thereby sending a *transverse* "whipped pulse" (vibration) out along the rope. See Figure 2.1(a). Imagine further that you can change the direction of the "whipped shape," quickly and randomly at your end, so that a person looking back along the rope toward you, sees the "vibration" occurring in all directions up and down, left to right, northeast to southwest, and so on, as shown in Figure 2.1(b).

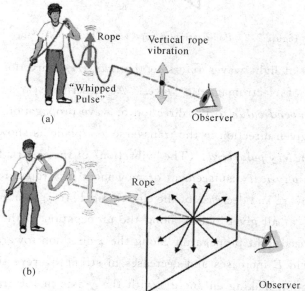

Figure 2.1 Rope waves and polarization

Optical Sensing and Measurement

In Figure 2.1, the rope wave is *linearly polarized*, that is, the rope vibrates in only one transverse direction vertically in the sketch shown. In Figure 2.1(b), the rope vibrations are in all transverse directions, so that the rope waves are said to be *unpolarized*.

Now imagine that the waves on the rope representing all possible directions of vibration as shown in Figure 2.1(b) are passed through a *picket fence*. Since the vertical slots of the fence pass only vertical vibrations, the many randomly oriented transverse vibrations incident on the picket fence emerge as only vertical vibrations, as depicted in Figure 2.2. In this example of transverse waves moving out along a rope, we see how we can with the help of a polarizing device, the picket fence in this case change unpolarized rope waves into polarized rope waves.

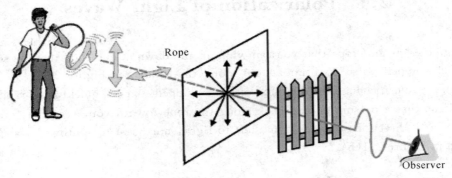

Figure 2.2 Polarization of rope waves by a picket fence

The polarization of light waves refers to the *transverse* direction of vibration of the electric field vector of electromagnetic waves. As described earlier, *transverse* means E-field vibrations *perpendicular* to the direction of wave propagation. If the electric field vector remains in a given direction in the transverse x-y plane as shown in Figure 2.3 the light is said to be *linearly polarized*. (The "vibration" of the electric field referred to here is not the same as a *physical* displacement or movement in a rope. Rather, the vibration here refers to an increase and decrease of the electric field strength occurring in a particular transverse direction. at all given points along the propagation of the wave.) Figure 2.3 shows linearly polarized light propagating along the z-direction toward an observer at the left. The electric field E increases and decreases in strength, reversing itself as shown, always along a direction making an angle θ with the y-axis in the transverse plane. The E-field components $E_x = E\sin\theta$ and $E_y = E\cos\theta$ are shown also in the figure.

Chapter 2 Polarization

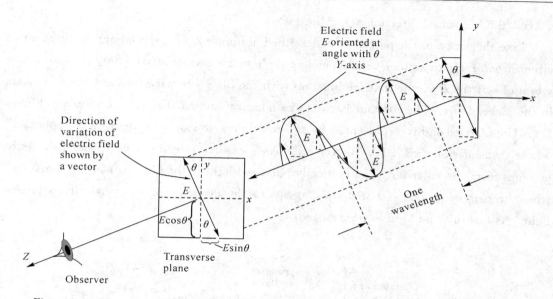

Figure 2.3 Linearly polarized light with transverse electric field E propagating along the z-axis

Table 2.1 Standard Symbols for Polarized Light

Viewing Position	Unpolarized	Vertically Polarized	Horizontally Polarized
Viewed head-on; beam coming toward viewer	✳	↕ ·	←·→
Viewed from the side; beam moving from left to right	↕·↕·↕· →	↕↕↕↕↕ →	•••• →

Table 2.1 lists the *symbols* used generally to indicate *unpolarized* light (**E**-vector vibrating randomly in all directions), *vertically polarized* light (**E**-vector vibrating in the vertical direction only), and *horizontally polarized* light (**E**-vector vibrating in the horizontal direction only). With reference to Figure 2.3, the vertical direction is along the

Optical Sensing and Measurement

y-axis, the horizontal direction along the x-axis.

Like the action of the picket fence described in Figure 2.2, a special optical filter called either a *polarizer* or an *analyzer* depending on how it's used transmits only the light wave vibrations of the **E**-vector that are lined up with the filter's *transmission axis* like the slats in the picket fence. The combined action of a polarizer and an analyzer are shown in Figure 2.4. Unpolarized light, represented by the multiple arrows, is incident on a "polarizer" whose transmission axis (TA) is vertical. As a result, only *vertically polarized* light emerges from the polarizer. The vertically polarized light is then incident on an "analyzer" whose transmission axis is horizontal, at 90° to the direction of the vertically polarized light. As a result, **no** light is transmitted.

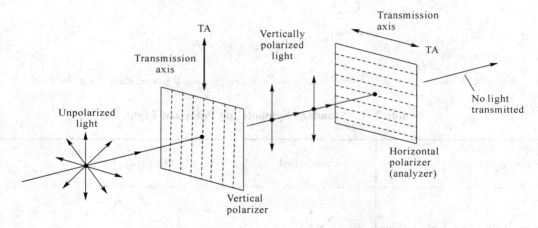

Figure 2.4 Effect of polarizers on unpolarized light

2.2 Law of Malus

When unpolarized light passes through a polarizer, the light intensity proportional to the square of its electric field strength is reduced, since only the E-field component along the transmission axis of the polarizer is passed. When linearly polarized light is directed through a polarizer and the direction of the E-field is at an angle θ to the transmission axis of the polarizer, the light intensity is likewise reduced. The reduction in intensity is expressed by the *law of Malus*, given in Equation (2.1).

$$I = I_0 \cos^2\theta \tag{2.1}$$

where I = intensity of light that is passed through the polarizer

I_0 = intensity of light that is incident on the polarizer

θ = angle between the transmission axis of the polarizer and the direction of the E-field

Application of the law of Malus is illustrated in Figure 2.5, where two polarizers are used to control the intensity of the transmitted light. The first polarizer changes the incident unpolarized light to linearly polarized light, represented by the vertical vector labeled E_0. The second polarizer, whose TA is at an angle θ with E_0, passes only the component $E_0 \cos\theta$, that is, the part of E_0 that lies along the direction of the transmission axis. Since the intensity goes as the square of the electric field, we see that I, the light intensity transmitted through polarizer 2, is equal to $(E_0 \cos\theta)^2$, or $I = E_0^2 \cos^2\theta$. Since E_0^2 is equal to I_0, we have demonstrated how the *law of Malus* ($I = I_0 \cos^2\theta$) comes about.

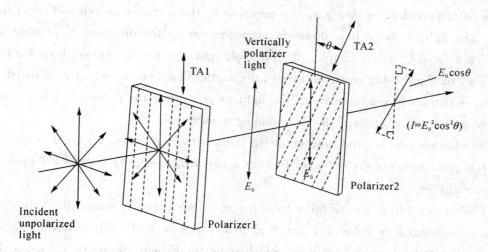

Figure 2.5 Controlling light intensity with a pair of polarizers

We can see that, by rotating polarizer 2 to change θ, we can vary the amount of light passed. Thus, if $\theta = 90°$ (TA of polarizer 1 is 90° to TA of polarizer 2) no light is passed, since $\cos 90° = 0$. If $\theta = 0°$ (TA of polarizer 1 is parallel to TA of polarizer 2) all of the light is passed, since $\cos 0° = 1$. For any other θ between 0° and 90°, an amount $I_0 \cos^2\theta$ is passed.

Optical Sensing and Measurement

2.3 Polarization by Reflection and Brewster's Angle

Unpolarized light, the light we normally see around us can be polarized through several methods. The polarizers and analyzers we have introduced above polarize by *selective absorption*. That is, we can prepare materials called *dichroic* polarizers that selectively *absorb* components of E-field vibrations along a given direction and largely *transmit* the components of the E-field vibration perpendicular to the absorption direction. The perpendicular (transmitting) direction defines the TA of the material. This phenomenon of selective absorption is what E. H. Land discovered in 1938 when he produced such a material, and called it *Polaroid*.

Polarization is produced also by the phenomenon of *scattering*. If light is incident on a collection of particles, as in a gas, the electrons in the particles absorb and reradiate the light. The light radiated in a direction perpendicular to the direction of propagation is partially polarized. For example, if you look into the north sky at dusk through a polarizer, the light being scattered toward the south toward you is partially polarized. You will see variations in the intensity of the light as you rotate the polarizer, confirming the state of partial polarization of the light coming toward you.

Another method of producing polarized light is by *reflection*. Figure 2.6 shows the *complete* polarization of the *reflected light* at a *particular angle of incidence B*, called the *Brewster angle*.

The refracted light on the other hand becomes only partially polarized. Note that the symbols introduced in Table 2.1 are used to keep track of the different components of polarization. One of these is the dot which indicates E-field vibrations perpendicular to both the light ray and the plane of incidence, that is, in and out of the paper. The other is an arrow indicating E-field vibrations **in** the plane of incidence **and perpendicular** to the ray of light. The reflected E-field coming off at Brewster's angle is totally polarized in a direction in and out of the paper, perpendicular to the reflected ray. This happens only at Brewster's angle, that particular angle of incidence for which the angle between the reflected and refracted rays, $B+\beta$, is exactly 90°. At the angle of incidence B, the E-field component cannot exist, for if it did it would be **along** the reflected ray, violating the requirement that E-field vibrations must always be transverse. that is, perpendicular to the direction of propagation. Thus, only the E-field component perpendicular to the plane of incidence is reflected.

Referring to Figure 2.6 and Snell's law at the Brewster angle of incidence, we can write:

$$n_1 \sin B = n_2 \sin \beta \tag{2.2}$$

Since $\beta + B = 90°$, $\beta = 90 - B$, which then allows us to write

$$n_1 \sin \beta = n_2 \sin(90 - B) = n_2 \cos B \tag{2.3}$$

or

$$\frac{\sin B}{\cos B} = \frac{n_2}{n_1} \tag{2.4}$$

and finally

$$\tan B = \frac{n_2}{n_1} \tag{2.5}$$

Equation (2.5) is an expression for Brewster's law. Knowing n_1 (the refractive index of the *incident* medium) and n_2 (the refractive index of the *refractive* medium), we can calculate the Brewster angle B. Shining light on a reflecting surface at this angle ensures complete polarization of the reflected ray. We make use of Equation (2.5) in example.

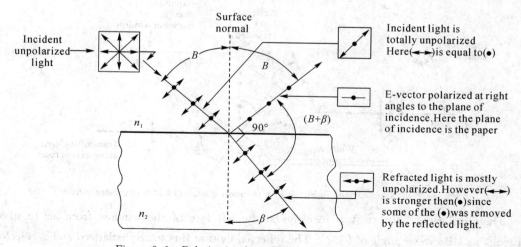

Figure 2.6 Polarization by reflection at Brewster's angle

Optical Sensing and Measurement

2.4 Brewster Windows in A Laser Cavity

Brewster windows are used in laser cavities to ensure that the laser light, after bouncing back and forth between the cavity mirrors, emerges as linearly polarized light. Figure 2.7 shows the general arrangement of the windows, thin slabs of glass with parallel sides mounted on the opposite edges of the gas laser tube, in this case a helium-neon gas laser.

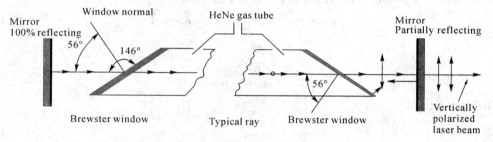

Figure 2.7 Brewster windows in a HeNe gas laser

As you can see, the light emerging is linearly polarized in a vertical direction. Why this is so is shown in detail in Figure 2.8. Based on Figure 2.6, Figure 2.8 shows that it is the refracted light, and not the reflected light, that is eventually linearly polarized.

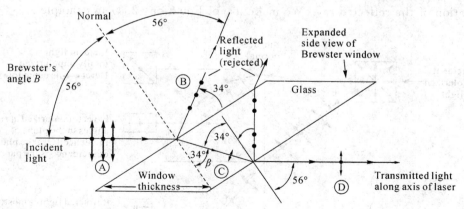

Figure 2.8 Unpolarized light passing through both faces at a Brewster' angle

The unpolarized light at A is incident on the left face of the window from air to glass, defining, as a Brewster angle of 56.3°. The reflected light at B is totally polarized and is *rejected*. The refracted (transmitted) light at C is now partially polarized since the reflected light has carried away part of the vibration perpendicular to the paper (shown by the dots). At the right face, the ray is incident again at a Brewster angle (34°) for a glass-to-air interface, as was shown in Figure 2.8. Here again, the reflected light, totally polarized, is *rejected*. The light transmitted through the window, shown at D, now has even less of the vibration perpendicular to the paper. After hundreds of such passes back and forth through the Brewster windows, as the laser light bounces between the cavity mirrors, the transmitted light is left with only the vertical polarization, as shown exiting the laser in Figure 2.8. And since all of the reflected light is removed (50% of the inital incident light) we see that 50% of the initial incident light *remains* in the refracted light, hence in the laser beam.

2.5 Polarization and Electromagnetic Effects

Light is an electromagnetic wave. The electric field of a light wave propagating in the z-direction is given by $\boldsymbol{E} = \boldsymbol{E}_0 \exp(kz - \omega t)$. The time-averaged intensity of the wave is $(1/(2\mu_0 c))|E_0|^2$. Light is a transverse wave. $\boldsymbol{E}$ is a vector lying in the plane perpendicular to z, $\boldsymbol{E} = (E_x, E_y)$. Polarized light is produced when the direction of $\boldsymbol{E}$ in the plane perpendicular to the direction of propagation is constrained in some fashion (see Figure 2.9).

The electric field vector $\boldsymbol{E}$ can always be resolved into two perpendicular components. The light is elliptically polarized, then the two components have a constant phase difference, and the tip of the electric field vector traces out an ellipse in the plane perpendicular to the direction of propagation.

$$E_x = E_{0x} \exp(i(kz - \omega t)), \quad E_y = E_{0y} \exp(i(kz - \omega t + \phi)) \tag{2.6}$$

Linearly polarized light is a special case of elliptically polarized light. If the light is linearly polarized, then the two components oscillate in phase, $E_x = E_{0x} \exp(i(kz - \omega t))$, $E_y = E_{0y} \cdot \exp(i(kz - \omega t))$, $\phi = 0$. The direction of $\boldsymbol{E}$ and the direction of propagation define a plane. The electric vector traces out a straight line. For example, $\boldsymbol{E} = E\boldsymbol{i} = E_{0x} \exp(i(kz - \omega t))\boldsymbol{i}$.

Circularly polarized light is also a special case of elliptically polarized light in which $E_{0x} = E_{0y}$ and the two components have a 90° phase difference and the electric field vector traces out a circle in the plane perpendicular to the direction of propagation. When viewed looking towards the source, a right circularly polarized beam has a field vector that describes a clockwise circle ($\phi = -\pi/2$), while left circularly polarized light has a field vector that describes a counter-clockwise circle ($\phi = \pi/2$).

Optical Sensing and Measurement

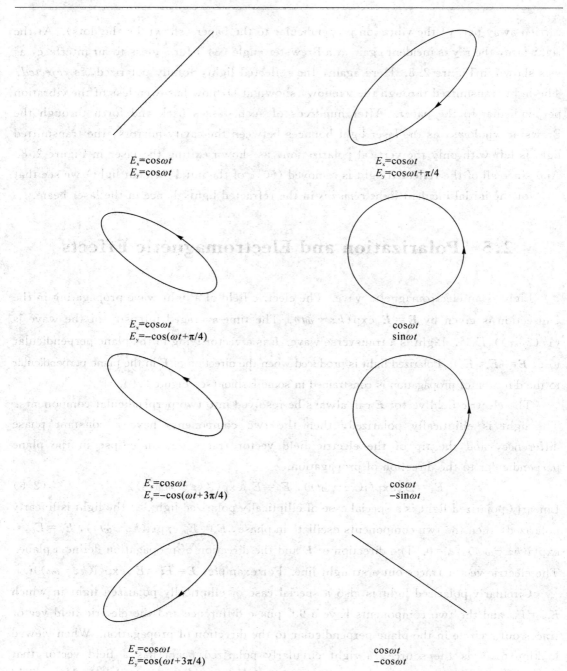

Figure 2.9 The polarized light

Let $E_x = E_{0x}\exp(i(kz-\omega t))$, $E_y = E_{0y}\exp(i(kz-\omega t+\phi))$.

- Linearly polarized light: $\phi = n\pi$, $n = 0, 1, 2, \ldots$
- Circularly polarized light: $E_{0x} = E_{0y}$, $\phi = n\pi/2$, $n = 1, 3, 5, \ldots$
- Elliptically polarized light: $\phi =$ arbitrary, but constant.
- Unpolarized light: ϕ, E_y, E_x are randomly varying on a timescale that is much shorter than that needed for observation.

Polarization Mechanisms

Dichroism: Certain naturally occurring crystalline materials have transmittance properties which depend on the polarization state of light. Dichroism is the selective absorption of one plane of polarization in preference to the other, orthogonal polarization during transmission through the material. The most common method of producing polarized light is to use polaroid material, made from chains of organic molecules, which are anisotropic in shape. Light transmitted is linearly polarized perpendicular to the direction of the chains. If linearly polarized light passes through polaroid material, then the transmitted intensity is given by $I_t = I_0 \cos^2\theta$, (Law of Malus), where θ is the angle between $\mathbf{E}$ and transmission direction.

Reflection: When unpolarized light is incident on a boundary between two dielectric surfaces, for example on an air-glass boundary, then the reflected and transmitted components are partially plane polarized. The Fresnel reflection coefficients are different for p- and s-polarized light. The reflected wave is 100% linearly polarized when the incident angle is equal to the Brewster angle θ_B. We then have $\tan\theta_B = n_2/n_1$.

- If a number of plates are stacked parallel to one another, and the entire pile is oriented at the Brewster angle, some of the s-polarized light will be reflected and all of the p-polarized light will be refracted at each interface. After one interface the refracted beam will be partially polarized, having lost some of its s-polarized component. If the stack contains many plates, then the refracted beam will have a high degree of polarization, since at each interface the same fraction of the remaining s-polarization is lost. This pile-of-plates polarization mechanism is used in many polarizing beam splitters, where many layers of dielectric thin film are laid onto the interior prism angle of the beam splitter.

Birefringence (double refraction): Certain crystalline substances have a refractive index which depends upon the state of incident polarization. Unpolarized light entering a birefringent crystal not along the optic axis of the crystal is split into beams which are refracted by different amounts.

A **linearly birefringent** crystal, such as calcite, will divide an entering beam of monochromatic light into two beams having orthogonal polarizations (see Figure 2.10). The beams will usually propagate in different directions and have different propagation speeds. Depending on whether the birefringent crystal is uniaxial or biaxial, there will be one or two directions within the crystal along which the beams will remain colinear and continue to propagate with the same speed. These directions are called the optic axes directions. If the crystal is a plane-parallel plate, and the optic axes directions are not collinear with the beam, the radiation will emerge as two separate, orthogonally polarized beams.

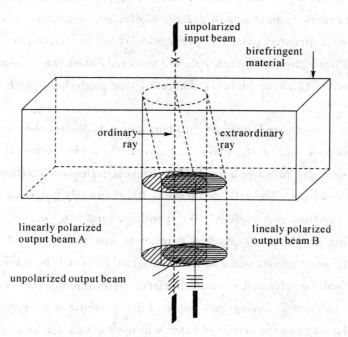

Figure 2.10 Linearly birefringent crystal

The two beams within the birefringent cystal are referred to as the ordinary and extraordinary ray, respectively (see Figure 2.11). The polarization of the extraordinary ray lies in the plane containing the direction of propagation and the optic axis, and the polarization of the ordinary ray is perpendicular to this plane.

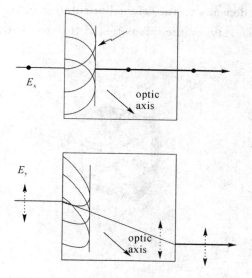

Figure 2.11 The extraordinary ray and the ordinary ray

The extraordinary ray violates both Snell's Law and the Law of Reflection. It is not necessarily confined to the plane of incidence. Its speed changes with direction. The index of refraction for the extraordinary ray is a continuous function of direction. The index of refraction for the ordinary ray is independent of direction. The two indices of refraction are equal only in the direction of an optic axis. When the ordinary index of refraction is plotted against wavelength, the dispersion curve for the ordinary ray is a single unique curve. The dispersion curve for the extraordinary ray is a family of curves with different curves for different directions. A ray normally incident on a birefringent crystalline surface will be divided into two rays at the boundary, unless it is in a special polarization state or unless the crystalline surface is perpendicular to an optic axis. The extraordinary ray will deviate from the incident direction while the ordinary ray will not. The ordinary ray index n_0 and the most extreme extraordinary ray index n_e are together known as the principal indices of refraction of the material. The direction of the lesser index is called the fast axis because light polarized in that direction has the higher speed.

If a beam of linearly polarized monochromatic light enters a birefringent crystal along a direction not parallel to the optical axis of the crystal, the beam will be divided into two separate beams. Each will be polarized at right angles to the other, and they will travel in different directions. The intensity of the original beam will be divided between the two new beams in a manner which depends on the original orientation of the electric field vector with respect to the crystal. The ratio or the intensities of the two orthogonally polarized beams can have any value.

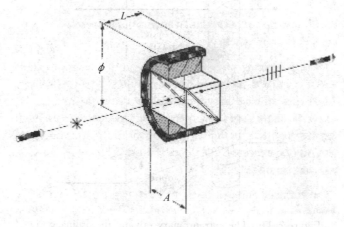

Figure 2.12   The Glan-Taylor polarizing prism

Birefringent crystal are used in many polarization devices. In some devices the difference in the refractive index is used to separate the rays and eliminate one of the polarization states, as in the various Glan-type polarizers.

In the Glan-Taylor polarizing prism shown above the rejected (ordinary) ray is absorbed by black mounting material in the prism housing(see Figure 2.12).

In other devices the changes in direction of propagation between the two rays is used to separate the incoming beam into two orthogonally polarized beams as in the Wollaston and Thompson beamsplitting prisms(see Figure 2.13).

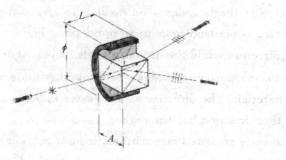

Figure 2.13   The Wollaston prism

Other device maintain the existing direction of propagation of the beam while separating the two polarized beam perpendicular to the direction of propagation into two parallel beams as in the beam displacer shown in Figure 2.14.

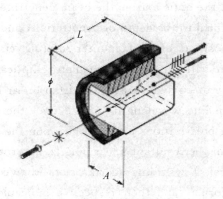

Figure 2.14 The beam displacer

A thin plate of birefringent crystal of thickness d cut parallel to the optic axis is known as a wave plate. Assume linearly polarized light is entering the wave plate normally. The components of **E** parallel and perpendicular to optic axis emerge with a phase difference δ between them given by $\delta = (2\pi d\lambda)\Delta n$.

A quarter-wave plate $\delta = \pi/2$ can be used to convert linearly polarized light to circularly polarized light. The incident linearly polarized light must be oriented at 45° to the wave plate's axes.

A half-wave plate $\delta = \pi$ can be used to rotate the plane of linearly polarized light. The angle of rotation is 2θ, where θ is the angle between the angle of polarization and the wave plate's fast axis.

If a plane polarized beam propagates down the optic axis of a material exhibiting **circular birefringence** it is resolved into two collinear circularly polarized beams, each propagating with a slightly different velocity.

In non-active crystals (linear birefingence), such as calcite there exist a direction in which the indices of refraction the ordinary (O) and the extraordinary (E) ray are exactly equal. In active crystals (circular birefringence), such as quartz, there is no such direction. The direction of the optical axis for active crystals is the direction in which the difference in the indices for the O and E ray is a minimum. If a quartz plate is cut such that its optic axis is normal to the surfaces of the plate, and a ray of linearly polarized light is incident parallel to the optical axis, the ray will be separated into two collinear, circularly polarized rays. The ordinary (O) and the extraordinary (E) rays will have opposite senses of circular polarization and will travel at different speeds. This causes the resultant plane of polarization to rotate about the optic axis as the beam penetrates the plate. The amount of rotation is directly proportional to the depth of penetration, and ultimately to the thickness of the plate. The superposition of two counter-rotating circular polarizations always produces linear polarization without any intermediate elliptical polarization states. This distinguishes optical activity or circular retardation from linear retardation. For crystalline quartz at the yellow sodium line wavelength (589.3 nm) the difference in the refractive indices for opposite circular polarizations propagating along the optic axis is only about one part in 10,000. The plane of the resultant linear polarization rotates at a rate of about 21.7 degrees per millimeter travel. Crystalline quartz exists in two distinct crystalline forms. Natural crystals of these forms, and their respective molecules, have shapes which are mirror images of each other. The plane of the resultant linear polarization rotates in opposite directions in these forms. The two crystalline forms are known as left-handed and right-handed quartz.

The polarization of the O-and E-rays in quartz rapidly changes from circular to elliptical even for directions which depart only slightly from the optical axis. For sodium light at an angle of only five degrees from the axis, the polarization ellipse has a major to minor axis ratio of 2.37. For the ellipse to be even approximately circular, much smaller angles are required. For this reason, devices which depend on circular polarization are effective only when highly collimated light propagates parallel to the optical axis direction.

Before discussing non-linear effects, let us look at a mathematical description of polarization. The electric field of any polarized beam propagating along the z-axis may be written as

$$E = E_x \mathbf{i} + E_y \mathbf{j} \tag{2.7}$$

where $E_x = A_x \exp(i(kz - \omega t + \phi_x))$, $E_y = A_y \exp(i(kz - \omega t + \phi_y))$

We can write the components as a column vector, which is called a **Jones vector**.

$$E = \begin{pmatrix} E_x \\ E_y \end{pmatrix} = \begin{pmatrix} A_x \exp(i(kz - \omega t + \phi_x)) \\ A_y \exp(i(kz - \omega t + \phi_y)) \end{pmatrix} \tag{2.8}$$

We may factor out and the dependence on z and t and just write

$$E = \begin{pmatrix} E_x \\ E_y \end{pmatrix} = \begin{pmatrix} A_x \exp(i\phi_x) \\ A_y \exp(i\phi_y) \end{pmatrix} \tag{2.9}$$

Chapter 2 Polarization

The intensity of the beam is proportional to $A_x^2 + A_y^2$. The most general Jones vector of a polarized beam propagating along the z-axis is given by the above equation.

The Jones vector for horizontally polarized light is given by

$$E = \begin{pmatrix} A_x \exp(i\phi_x) \\ 0 \end{pmatrix} = A_x \exp(i\phi_x) \begin{pmatrix} 1 \\ 0 \end{pmatrix} \tag{2.10}$$

Similarly, the Jones vector for vertically polarized light is given by

$$E = \begin{pmatrix} 0 \\ A_y \exp(i\phi_y) \end{pmatrix} = A_y \exp(i\phi_y) \begin{pmatrix} 0 \\ 1 \end{pmatrix} \tag{2.11}$$

The normalized Jones vector for light polarized at 45° is given by

$$E = \begin{pmatrix} A\exp(i\phi) \\ A\exp(i\phi) \end{pmatrix} = \frac{1}{\sqrt{2}} \begin{pmatrix} i \\ 1 \end{pmatrix}, \tag{2.12}$$

where $A_x = A_y = A = 1$ and $\phi_x = \phi_y = \phi = \pi/4$.

The **normalized** Jones vector for right-hand circularly polarized light is given by

$$E = \begin{pmatrix} A\exp(i\phi) \\ A\exp(i(\phi-\pi/2)) \end{pmatrix} = \frac{1}{\sqrt{2}} \begin{pmatrix} 1 \\ -i \end{pmatrix}. \tag{2.13}$$

For the normalized vectors we have

$$\frac{1}{\sqrt{2}} \begin{pmatrix} 1 \\ -i \end{pmatrix} = \frac{1}{\sqrt{2}} \begin{pmatrix} i \\ 1 \end{pmatrix}. \tag{2.14}$$

If a polarized beam with field vector E is incident on a polarization-changing medium such as a polarizer or a wave plate, and the result is a beam in another polarization state given by E' with $E'_x = m_{11}E_x + m_{12}E_y$, $E'_y = m_{21}E_x + m_{22}E_y$, then we may write

$$\begin{pmatrix} E_x' \\ E_y' \end{pmatrix} = \begin{pmatrix} m_{11} & m_{12} \\ m_{21} & m_{22} \end{pmatrix} \begin{pmatrix} E_x \\ E_y \end{pmatrix}, \tag{2.15}$$

where the 2 by 2 transformation matrix is called the **Jones matrix**. The table (see Table 2.2) below lists the Jones matrices for common optical elements.

Table 2.2   The Jones Matrices for Common Optical Elements

Optical Element	Jones Matrix
Horizontal linear polarizer	$\begin{pmatrix} 1 & 0 \\ 0 & 0 \end{pmatrix}$
Vertical linear polarizer	$\begin{pmatrix} 0 & 0 \\ 0 & 1 \end{pmatrix}$

Continued

Optical Element	Jones Matrix
Linear polarizer at	$\begin{pmatrix} \cos^2\theta & \cos\theta\sin\theta \\ \cos\theta\sin\theta & \sin^2\theta \end{pmatrix}$
Quarter wave plate (fast axis vertical)	$e^{i\pi/4} \begin{pmatrix} 1 & 0 \\ 0 & -i \end{pmatrix}$
Quarter wave plate (fast axis horizontal)	$e^{i\pi/4} \begin{pmatrix} 1 & 0 \\ 0 & i \end{pmatrix}$

If we require the Jones matrix for an optical element which has been rotated through an angle θ with respect to the direction given in the table above, we must multiply the above matrix by the usual matrices for rotation.

$$M(\theta) = R(\theta) \, M \, R(-\theta) \tag{2.16}$$

where

$$R = \begin{pmatrix} \cos\theta & -\sin\theta \\ \sin\theta & \cos\theta \end{pmatrix}. \tag{2.17}$$

To find the Jones matrix for a sequence of polarization transformations, for example a linear polarizer followed by a quarter wave plate, we simply multiply the individual Jones matrices together in the correct order. If an incident beam of light with field vector $\boldsymbol{E}$ passes through a sequence of four polarizing elements, M_1 followed by M_2, M_3 and M_4, then the resultant field vector $\boldsymbol{E}'$ is given by

$$\boldsymbol{E}' = M_4 M_3 M_2 M_1 \boldsymbol{E} \tag{2.18}$$

The eigenvectors of the Jones matrix M of an optical device correspond to the polarization states which propagate through the optical device represented by M unchanged. The beam enters and emerges in the same polarization state. The eigenvalues, which in general are complex numbers, tell us about changes in amplitude and phase produced by the optical device.

Chapter 3　Fiber Optics

3.1　Geometrical Optics

An optical wave guide is a structure that "guides" a light wave by constraining it to travel along a certain desired path. Usually the light is guided by total internal reflection (TIR). TIR occurs when light is incident on a dielectric interface at an angle greater than the critical angle θ_c.

A wave guide(see Figure 3.1) traps light by surrounding a guiding region, called the core, made from a material with index of refraction n_{core}, with a material called the cladding, made from a material with index of refraction $n_{cladding} < n_{core}$. Light entering is trapped as long as $\sin \theta > n_{cladding}/n_{ncore}$.

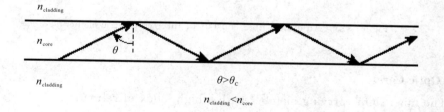

Figure 3.1　The wave guide

Light can be guided by planar or rectangular wave guides, or by optical fibers(see Figure 3.2). An optical fiber consists of three concentric elements, the core, the cladding and the outer coating, often called the buffer(see Figure 3.3). The core is usually made of glass or plastic. The core is the light-carrying portion of the fiber. The cladding surrounds the core. The cladding is made of a material with a slightly lower index of refraction than the core. This difference in the indices causes total internal reflection to occur at the core-cladding boundary along the length of the fiber. Light is transmitted down the fiber and does not escape through the sides of the fiber.

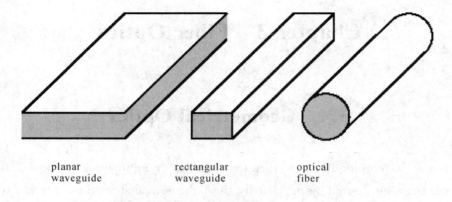

Figure 3.2   The optical fiber

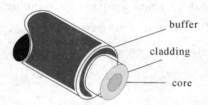

Figure 3.3   An optical fiber consists of three concentric elements

- **Fiber Optic Core:**
    • The inner light-carrying member with a high index of refraction.
- **Cladding:**
    • The middle layer, which serves to confine the light to the core. It has a lower index of refraction.
- **Buffer:**
    • The outer layer, which serves as a "shock absorber" to protect the core and cladding from damage. The coating usually comprises one or more coats of a plastic material to protect the fiber from the physical environment. Sometimes metallic sheaths are added to the coating for further physical protection.

Light injected into the fiber optic core and striking the core-to-cladding interface at an angle greater than the critical angle is reflected back into the core. Since the angles of incidence and reflection are equal, the light ray continues to zig-zag down the length of the fiber. The light is trapped within the core(see Figure 3.4). Light striking the interface at less than the critical angle passes into the cladding and is lost.

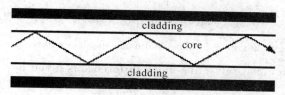

Figure 3.4   The light is trapped within the core

Optical fibers usually are specified by their size. Usually the outer diameter of the core, the cladding and the buffer are specified. For example, 62.5/120/250 refers to a fiber with a 62.5 μm diameter core, a 120 μm diameter cladding and a 0.25 mm outer coating diameter.

3.2   Wave Optics

The laws governing the propagation of light in optical fibers are Maxwell's equations. When information about the material constants, such as the refractive indices, and the boundary conditions for the cylindrical geometry of core and cladding is incorporated into the equations, they can be combined to produce a wave equation that can be solved for those electromagnetic field distributions that will propagate through the fiber. These allowed distributions of the electromagnetic field across the fiber are referred to as the modes of the fiber. They are similar to the modes found in microwave cavities and laser cavities. When the diameter of the core is large compared to the wavelength of the light propagating through the fiber, then the number of allowed modes becomes large and ray optics gives an adequate description of light propagation in fibers. Those fibers are called **multimode fibers**.

Multimode fibers for which the refractive index of the core is a constant and the index changes abruptly at the core-cladding interface are called **step-index fibers**(see Figure 3.5).

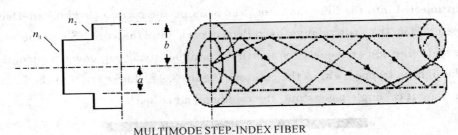

MULTIMODE STEP-INDEX FIBER

Figure 3.5 The multimode step-index fiber

For such fibers the fractional refractive index difference is given by

$$\Delta = \frac{n_{core} - n_{cladding}}{n_{core}} \tag{3.1}$$

The cone angle θ_{cone} of the cone of light that will be accepted by an optical fiber with a fractional index difference Δ is given by

$$n_i \sin\theta_{cone} = (n_{core}^2 - n_{cladding}^2)^{1/2} \tag{3.2}$$

Here n_i is the index of refraction of the material from which the light is entering the fiber (see Figure 3.6).

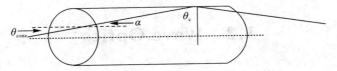

Figure 3.6 The cone angle θ_{cone} of the cone of light

$$\sin\theta_c = \frac{n_{cladding}}{n_{core}} = \cos\alpha = (1 - \sin^2\alpha)^{1/2}.$$

$$n_{core} \times \sin\alpha = (n_{core}^2 - n_{cladding}^2)^{1/2}.$$

Snell's law: $n_i \sin\theta_{cone} = n_{core} \times \sin\alpha$

The numerical aperture (NA) is the measure of of how much light can be collected by an optical system. For a fiber, it is defined as n_i times the sine of the maximum angle at which light rays can enter the fiber and be conducted down the fiber. It is given by $NA = (n_{core}^2 - n_{cladding}^2)^{1/2}$. When $\Delta \ll 1$, this can be approximated by

$$NA = ((n_{core} - n_{cladding})(n_{core} + n_{cladding}))^{1/2} = (2n_{core}^2\Delta)^{1/2} = n_{core}(2\Delta)^{1/2}.$$

The condition $\Delta \ll 1$ is referred to as the weakly-guiding approximation.

Each mode in a step-index multimode fiber is associated with a different entrance angle. Each mode therefore travels along a different path through the fiber. Different propagating modes have different group velocities. As an optical pulse travels down a multimode fiber, the pulse begins to spread (see Figure 3.7). Pulses that enter separated from each other will eventually overlap each other. This limits both the bandwidth of a multimode fiber and the distance over which it can transport data.

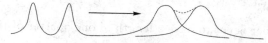

Figure 3.7 The pulse spread

Bandwidth measures the data-carrying capacity of an optical fiber. It is expressed as the product of the data frequency and the distance over which data can be transmitted at that frequency. For example a fiber with a bandwidth of 400 MHz km can transmit data at a rate of 400 MHz for 1 km or at a rate of 20 MHz for 20 km. Step-index fibers have a typical bandwidth of 20 MHz km.

Step-index fibers are available with core diameters of 100 to 1000 μm. They are well suited to applications requiring high-power densities, such delivering laser power for medical and industrial applications.

The smearing of the pulses in step-index fibers scan be reduced through the use of graded-index or single-mode fibers.

The core in a **graded-index fiber** (see Figure 3.8) has an index of refraction that decreases as the radial distance from the center of the core increases. As a result, the light travels faster near the edge of the core than near the center. Different modes therefore travel in curved paths with nearly equal travel times. This greatly reduces modal dispersion in the fiber. Graded-index fibers therefore have bandwidths which are significantly greater than step-index fibers. Typical core diameters of graded-index fibers are 50, 62.5 and 100 μm. Graded-index fibers are often used in medium-range communications applications, such as local area networks. Graded-index fibers have a typical bandwidth of 500 MHz km at $\lambda=1300$ nm and 160 MHz km at $\lambda=850$ nm.

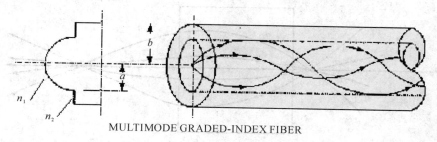

MULTIMODE GRADED-INDEX FIBER

Figure 3.8 The Graded-index fiber

Optical Sensing and Measurement

A fan of rays injected into a graded-index fiber is brought back into focus, before it diverges again. A ray will travel along an approximately sinusoidal path. The wavelength of this sinusoidal path is called the pitch of the fiber. The pitch is determined by Δ, the fractional index difference.

If a graded-index fiber is cut to have a length of one quarter of the pitch of the fiber, it can serve as an extremely compact lens, called a **GRIN lens**.

Light exiting a fiber can be collimated into a parallel beam when the output end of the fiber is connected to the GRIN lens. Because its properties are set by its length, this graded-index lens is referred to as a quarter-pitch or 0.25 pitch lens(see Figure 3.9).

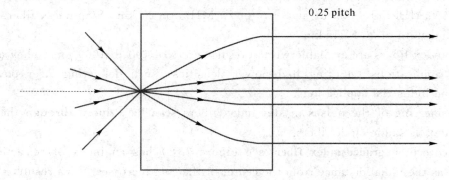

Figure 3.9　The GRIN lens(0.25 pitch lens)

Focusing of the fiber output onto a small detector or focusing of the output of a source onto the core of a fiber can be accomplishing by increasing the length of the GRIN lens to 0.29 pitch(see Figure 3.10). Then the source can be moved back from the lens and the transmitted light can be refocused at some point beyond the lens. Such an arrangement is useful for coupling sources to fibers and fibers to detectors.

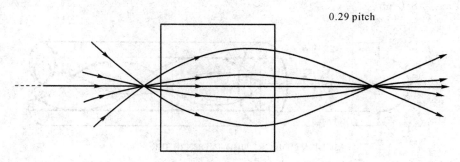

Figure 3.10　The GRIN lens (0.29 pitch lens)

The modes that propagate in a fiber are found by solving Maxwell's equation for the electric field of the light in the fiber in cylindrical coordinates. Solutions which are harmonic in space and time, are of the form

$$E(r, \phi, z) = f(r) \cos(\omega t - \beta z + \chi) \cos(q\phi) \qquad (3.3)$$

where ω is the angular frequency of light and β is the propagation constant. Here z is the direction of propagation, and q is an integer.

The group velocity of the mode is β/ω. It is important to make the distinction between the magnitude of the wave vector $\mathbf{k}$, and the magnitude of propagation constant β. In the ray approximation, β is the z-component of $\mathbf{k}$.

The normalized wave number, or V-number of a fiber is defined as $V = k_f a$ NA. Here k_f is the free space wave number, $2\pi/\lambda_0$, a is the radius of the core, and NA is the numerical aperture of the fiber. Many fiber parameters can be expressed in terms of V. For example, the number of guided modes n in a step-index multimode fiber is given by $V^2/2$ for $n \gg 1$, and a step index fiber becomes single-mode for a given wavelength when $V < 2.405$.

In the weakly-guiding approximation ($\Delta \ll 1$), the modes propagating in the fiber are linearly polarized (LP) modes characterized by two subscripts, m and n. The first subscript m, gives the number of azimuthal, or angular, nodes in the electric field distribution. The second subscript n, gives the number of radial nodes. Output patterns are symmetric about the center of the beam and show bright regions separated by dark regions (the nodes that determine the order numbers m and n). The zero field at the outer edge of the field distribution is counted as a node.

When the V number is less than 2.405 only the LP_{01} mode propagates. When the V number is greater than 2.405, the next linearly-polarized mode can be supported by the fiber, so that both the LP_{01} and LP_{11}, modes will propagate (see Figure 3.11).

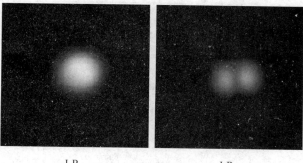

LP_{01} LP_{11}

Figure 3.11 The LP_{01} and LP_{11} modes

Multimode fibers used for telecommunications have V-numbers between ~50 or 150. A large number of modes are supported by these fibers. The amount of light carried by each mode is determined by the launch conditions. The attenuation of some large-angle modes is much higher than that of other modes, but after the light has propagated a considerable distance, a stable mode distribution develops. To generate a stable mode distribution even with only a short length of fiber, mode filtering is accomplished through **mode scrambling**.

A series of bends is introduced into the fiber (see Figure 3.12). These bends couple out the light in the large-angle modes with the high attenuation and distribute the remaining light among the other guided modes. Mode scrambling permits repeatable, accurate measurements of fiber attenuation to be made in the laboratory, even with short lengths of fiber.

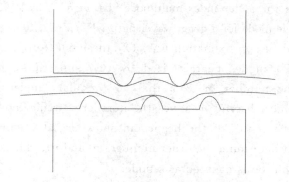

Figure 3.12 A series of bends is introduced into the fiber

Only the fundamental zero-order mode is transmitted in a **single mode fiber** (see Figure 3.13).

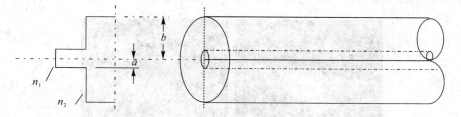

Figure 3.13 A single mode fiber

Because the single-mode fiber propagates only the fundamental mode, modal dispersion, the primary cause of pulse overlap, is eliminated. Thus, the bandwidth of a single-mode fiber is much higher than that of a multimode fiber. Pulses can be transmitted much closer together in time without overlap. Because of this higher bandwidth, single-mode fibers are used in all modern long-range communication systems. Typical core diameters are between 5 and 10 μm. Single-mode fibers have a typical bandwidth of 100 GHz km. Figure 3.14 shows light emerging from a multi-mode fiber and a single-mode fiber.

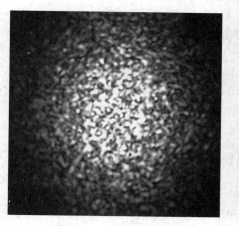 

(a) Light emerging from a multi-mode fiber   (b) Light emerging from a single-mode fiber

Figure 3.14 Light emerging from a multi-mode fiber and a single-mode fiber

Signals lose strength as they propagate through the fiber. This is known as beam attenuation. **Attenuation** is measured in decibels (dB).

$$A(\text{dB}) = 10 \lg\left(\frac{P_{\text{in}}}{P_{\text{out}}}\right) \quad \text{or} \quad 10^{(A/10)} = \frac{P_{\text{in}}}{P_{\text{out}}}. \quad P_{\text{out}} = 10^{-(A/10)} P_{\text{in}}$$

P_{in} and P_{out} refer to the optical power going in and coming out of the fiber. The table below shows the power typically lost in a fiber for several values of attenuation in decibels.

Attenuation (dB)	Power Loss (%)
10	90
3	50
0.1	2

The attenuation of an optical fiber is wavelength dependent. Attenuation is usually expressed in dB/km at a specific wavelength. Typical values range from 10 dB/km for step-index fibers at 850 nm to a few tenths of a dB/km for single-mode fibers at 1550 nm. There are several causes of attenuation in an optical fiber.

Optical Sensing and Measurement

• Rayleigh Scattering—Microscopic-scale variations in the index of refraction of the core material can cause considerable scatter in the beam leading to substantial losses of optical power.

• Absorption—Current manufacturing methods have reduced absorption caused by impurities to very low levels.

• Bending—Manufacturing methods can produce minute bends in the fiber geometry. Sometimes these bends will be great enough to cause the light within the core to hit the core/cladding interface at less than the critical angle so that light is lost into the cladding material. This also can occur when the fiber is bent in a tight radius. Bend sensitivity is usually expressed in terms of dB/km loss for a particular bend radius and wavelength.

3.3 Optical Connectors

Connectors are used to mate a fiber to another fiber or to equipment. Good coupling efficiency requires precise positioning of the fiber. Connectors are used when one expects that the connection must occasionally be broken.

Optical connectors are similar to their electrical counterparts in function and outward appearance. They must, however, be high precision devices. A connectors must center the fiber so that its light gathering core lies directly over and in line with a light source or another fiber to a tolerance of a few ten thousandths of an inch.

There are many different types of optical connectors in use today (see Figure 3.15).

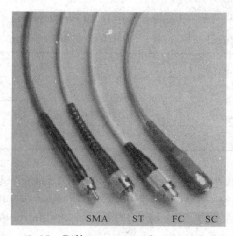

Figure 3.15 Different types of optical connectors

- The **SMA connector** was developed before the invention of single-mode fiber. Due to its stainless steel structure and low-precision, threaded fiber locking mechanism, this connector is used mainly in applications requiring the coupling of high-power laser beams into large-core, multimode fibers. Typical applications include laser beam delivery systems in medical and industrial applications. The typical insertion loss of an SMA connector is greater than 1 dB.

- The most popular type of multimode connector in use today is the **ST connector**. Initially developed by AT&T for telecommunications purposes, this connector uses a twist lock type design. Its high-precision, ceramic ferrule allows its use with both multimode and single-mode fibers. A typical mated pair of ST connectors will exhibit less than 1 dB (20%) of loss and does not require alignment sleeves or similar devices. The inclusion of an "anti-rotation tab" assures that every time the connectors are mated, the fibers always return to the same rotational position assuring constant, uniform performance.

- The **FC connector** has become the connector of choice for single-mode fibers, and is mainly used in fiber-optic instruments, SM fiber optic components, and in high-speed fiber optic communication links. This high-precision, ceramic ferrule connector is equipped with an anti-rotation key, reducing fiber end-face damage and rotational alignment sensitivity of the fiber. The key is also used for repeatable alignment of fibers in the optimal, minimal-loss position. The typical insertion loss of the FC connector is around 0.3 dB.

- The **SC connector** is becoming increasingly popular in single-mode fiber optic telecom and analog CATV, field deployed links. The high-precision, ceramic ferrule construction is optimal for aligning single-mode optical fibers. The connector's outer, square profile combined with its push-pull coupling mechanism, allow for greater connector packaging density in instruments and patch panels. The keyed outer body prevents rotational sensitivity and fiber end-face damage. The typical insertion loss of the SC connector is around 0.3 dB.

Splices are permanent connections between two fibers made by arc-welding the fibers together (fusion splicing) or gluing them together (mechanical splicing.) Both splices are capable of splice losses in the range of 0.15 dB (3%) to 0.1 dB (2%).

- In a mechanical splice, the ends of two pieces of fiber are cleaned and stripped, then carefully butted together and aligned using a mechanical assembly. A gel is used at the point of contact to reduce light reflection and keep the splice loss at a minimum. The ends of the fiber are held together by friction or compression, and the splice assembly features a locking mechanism so that the fibers remained aligned.

• A fusion splice, by contrast, involves actually melting (fusing) together the ends of two pieces of fiber. The result is a continuous fiber without a break. Fusion splices require special expensive splicing equipment but can be performed very quickly, so the cost becomes reasonable if done in quantity. As fusion splices are fragile, mechanical devices are usually employed to protect them.

3.4 Optical Fiber Measurements

Various laboratory measurements are routinely performed on telecommunication fibers to test their performance as components of fiber-optics communication systems. Some of these measurements are listed below.

Attenuation is the loss of optical power as a result of absorption, scattering, bending, and other loss mechanisms as the light travels through the fiber. The total attenuation is a function of the wavelength of the light. The total attenuation A between two arbitrary points X and Y on the fiber is $A(\text{dB}) = 10 \lg (P_x/P_y)$. P_x is the power output at point X. P_y is the power output at point Y. Point X is assumed to be closer to the optical source than point Y. The attenuation coefficient or attenuation rate α is given by $\alpha(\text{dB/km}) = A/L$. Here L is the distance between points X and Y.

The **cutback method** is often used for measuring the total attenuation of an optical fiber. The cutback method involves comparing the optical power transmitted through a long piece of fiber to the power transmitted through a very short piece of the fiber. The cutback method requires that a test fiber of known length L be cut back to a length of approximate 2 m. It requires access to both ends of the fiber. The cutback method begins by measuring the output power P_y of the test fiber of known length L. Without disturbing the input conditions, the test fiber is cut back to a length of approximate 2 m. The output power P_x of the short test fiber is then measured and the fiber attenuation A and the attenuation coefficient α are calculated.

Different launch conditions can lead to different results. For multimode fiber, the distribution of power among the modes of the fiber must be controlled. This is accomplished by controlling the launch spot size, i.e. the area of the fiber face illuminated by the light beam, and the angular distribution of the light beam.

When the launch spot size is smaller than the area of the fiber face and the numerical aperture NA of the input radiation is smaller than the NA of the fiber, the fiber is said to be underfilled(see Figure 3.16). Most of the optical power is concentrated in the center of the fiber and mainly low-order modes are excited.

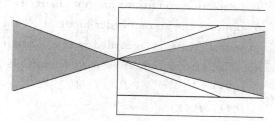

Figure 3.16　Underfilled launch conditions

When the launch spot size is larger than the area of the fiber face and the numerical aperture NA of the input radiation is larger than the NA of the fiber, the fiber is said to be overfilled(see Figure 3.17). Light that falls outside the fiber core and light incident at angles greater than the angle of acceptance of the fiber core is lost. Overfilling the fiber excites both low-order and high-order modes.

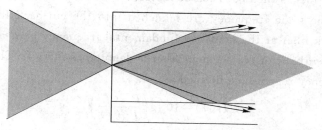

Figure 3.17　Overfilled launch conditions

Launch conditions affect the results of multimode fiber attenuation measurements. If too much power is launched into high-order modes, the high-order-mode power loss will dominate the attenuation results. Generally, fiber attenuation measurements are performed using underfilled launch conditions.

The **cutoff wavelength** of a single mode fiber is the wavelength above which the fiber propagates only the fundamental mode. We need $V = k_f a\text{NA} = 2\pi a \text{NA}/\lambda_0 < 2.405$. The cutoff wavelength of a single mode fiber is a function of the fiber's radius of curvature. Measuring the cutoff wavelength involves comparing the transmitted power from a test fiber with that of a reference fiber as a function of wavelengths.

Optical Sensing and Measurement

The test fiber is loosely supported in a single-turn with a constant radius of 140 mm. The transmitted signal power $P_s(\lambda)$ is recorded while scanning the wavelength in increments of 10 nm or less over the expected cutoff wavelength. The launch and detection conditions are not changed while scanning over the range of wavelengths. For the reference power measurement the launch and detection conditions are not changed, but the fiber is bent to a radius of 30 mm or less to suppress the second-order mode at all the scanned wavelengths. The transmitted signal power $P_r(\lambda)$ is recorded while scanning the over the same wavelength range as before. The attenuation at each wavelength is calculated.

$$P_s(\lambda) = 10^{-A_s(\lambda)/10}, \quad P_r(\lambda) = 10^{-A_r(\lambda)/10}$$
$$P_s(\lambda)/P_r(\lambda) = 10^{-(A_t(\lambda)-A_r(\lambda))/10} = 10^{A_d(\lambda)/10}$$

The attenuation at each wavelength is calculated. $A_d(\lambda)(\text{dB}) = 10 \lg(P_s(\lambda)/P_r(\lambda))$. The longest wavelength at which $A_d(\lambda)/\text{km}$ is equal to 0.1 dB is the fiber cutoff wavelength.

The modal **bandwidth** of a multimode optical fiber can be measured by measuring the power transfer function $H(f)$ of the fiber at the band frequency (f). Signals of varying frequencies (f) are launched into the test fiber and the power exiting the fiber at the launched fundamental frequency is measured. This optical output power is denoted as $P_{\text{out}}(f)$. The test fiber is then cut back or replaced with a short length of fiber of the same type. Signals of the same frequency are launched into the cut-back fiber and the power exiting the cut-back fiber at the launched fundamental frequency is measured. The optical power exiting the cutback or replacement fiber is denoted as $P_{\text{in}}(f)$. The magnitude of the optical fiber frequency response is defined as

$$H(f) = 10 \lg\left(\frac{P_{\text{out}}(f)}{P_{\text{in}}(f)}\right)$$

or, if the launch conditions for the two experiments are not exactly the same,

$$H(f) = 10 \lg\left(\frac{P_{\text{out}}(f) P_{\text{in}}(0)}{P_{\text{in}}(f) P_{\text{out}}(0)}\right)$$

The fiber bandwidth is defined as the frequency at which the magnitude of the fiber frequency response has decreased to one-half its zero-frequency value and $H(f) = -3$. This frequency is called the -3 decibel (dB) optical power frequency (f_{3dB}) and referred to as the fiber bandwidth. Bandwidth is normally given in units of megahertz-kilometers (MHz~km). Converting to a unit length assists in the analysis and comparison of optical fiber performance.

Chromatic dispersion occurs because the index of refraction is a function of wavelength and different wavelengths of light travel through the fiber at different speeds. The chromatic dispersion of multimode graded-index and single mode fibers is obtained by measuring the time it takes pulses of light with different wavelengths to travel through a long piece of fiber. These measurements are made using multi-wavelengths sources such as wavelength-selectable lasers or multiple sources of different wavelengths.

To make **fiber geometry** measurements, the input end of the fiber is overfilled and mode filtered. The output end of the fiber is viewed with a video camera. The image from the video camera is sent to a computer for digital analysis. The computer analyzes the image to identify the edges of the core and cladding. The **cladding diameter** is defined as the average diameter of the cladding. The **core diameter** is defined as the average diameter of the core. **Cladding noncircularity**, or ellipticity, is the difference between the smallest radius of the cladding and the largest radius of the cladding divided by the average cladding radius. **Core noncircularity** is the difference between the smallest core radius and the largest core radius divided by the average core radius. Core noncircularity is measured on multimode fibers only. The **core-cladding concentricity error** for multimode fibers is the distance between the core and cladding centers divided by the core diameter. The core-cladding concentricity error for single mode fibers is defined as the distance between the core and cladding centers.

The **core diameter** is measured by measuring the power distribution in the near-field region close to the fiber-end face, when the distance between the fiber-end face and detector is in the micrometers range. The core diameter (d) is defined as the diameter at which the intensity is 2.5 percent of the maximum intensity.

The **numerical aperture** (NA) is a measurement of the ability of an optical fiber to capture light. It is determined by measuring the far-field power distribution in the region far from the fiber-end face. The emitted power per unit area is recorded as a function of the angle some distance away from the fiber-end face. The distance between the fiber-end face and detector in the far-field region is in the centimeters (cm) range for multimode fibers and millimeters (mm) range for single mode fibers. The fiber NA is defined by the 5 percent or 0.05 intensity level. This 0.05 intensity level intersects the normalized curve at scan angles θ_A and θ_B. The fiber NA is defined as $NA = \sin((\theta_A - \theta_B)/2)$.

The **mode field diameter** (MFD) of a single mode fiber is related to the spot size of the fundamental mode. This spot has a mode field radius r_0. The mode field diameter is equal to $2r_0$. Single mode fibers with large mode field diameters are more sensitive to fiber bending. Single mode fibers with small mode field diameters show higher coupling losses at connections. The mode field diameter of a single mode fiber can be measured by measuring the output far-field power distribution of the single mode fiber using a set of apertures of various sizes.

Insertion loss is composed of the connection coupling loss and additional losses in the fiber following a connection. In multimode fiber, fiber joints can increase fiber attenuation following the joint by disturbing the mode power distribution. In single mode fibers, fiber joints can cause the second-order mode to propagate in the fiber following the joint. To measure insertion loss, power measurements are made on an optical fiber before the joint is inserted and after the joint is inserted. Initial power measurements at the detector (P_0) and at the source monitoring equipment (P_{M0}) are taken before inserting the interconnecting device into the test setup. The test fiber is then cut and the device is inserted. After insertion the power at the detector (P_1) and at the source monitoring equipment (P_{M1}) is measured again. The insertion loss is calculated as

$$\text{insertion loss (dB)} = 10 \lg\left(\left(\frac{P_1}{P_0}\right)\left(\frac{P_{M0}}{P_{M1}}\right)\right)$$

Reflection occurs at optical fiber connections. The **reflectance** R is the fraction of the incident intensity that is reflected back into the source fiber at the point of the connection. The **return loss** is defined as

$$\text{return loss} = -10 \lg R$$

The reflectance R is measured using an optical source connected to one input of a 2×2 fiber optic coupler. Light is launched into the component under test through the fiber optic coupler. The light reflected from the component under test is transmitted back through the fiber optic coupler to a detector connected to the other input port.

3.5 Fiber-Optic Communication

A fiber optic data link has three basic functions (see Figure 3.18). It must convert an electrical input signal to an optical signal, send the optical signal over an optical fiber, and then convert the optical signal back to an electrical signal.

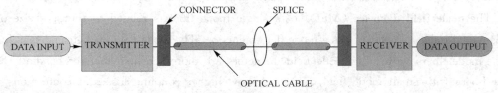

Figure 3.18 A fiber optic data link

Chapter 3 Fiber Optics

The **transmitter** converts the electrical input signal to an optical signal. Its **drive circuit** varies the current flow through the light source, which in turn varies the irradiance of the source. The process of varying the irradiance as a function of time is called **modulation**.

Analog modulation consists of changing the light level in a continuous manner. The performance of a system using analog modulation is limited by random noise in the system, either in the detector, which converts the modulated light signal back into an electrical signal, or in the system itself. Noise determines the smallest signal that can be transmitted and how faithful the reproduced signal is to the original signal.

In a system using **digital modulation** information is encoded into a series of pulses, separated by spaces. The absence or presence of a pulse at some point in the data stream represents one bit of information. Faithful reproduction of signal intensity is not required. Pulses must only be transmitted with sufficient power for the detector to determine the presence or absence of a pulse. This makes a system using digital modulation superior when sources of noise are present. Performance in digital systems is given in terms of the bit error rate, the average ratio of the number of errors to the number of transmitted pulses. State-of-the-art systems have bit error rates of less than 10^{-9}.

Fiber-optic telecommunication systems use pulse-code modulation. Information is transmitted as a series of pulses. The digital pulse-code-modulated signal is coupled into a fiber. The fiber end is positioned by a connector to maximize the input power. Semiconductor lasers are well-suited for use in a fiber-optic communication system. Their size and shape allows for efficient coupling of light into the small-diameter core of an optical fiber. Modern $Al_xGa_{1-x}As$ lasers operate continuously at mW power levels. Their output can be modulated easily at frequencies into the GHz range by modulating the output of an electric power supply.

Many existing systems use $Al_xGA_{1-x}As$ source lasers or LEDs operating near 0.85 μm. Such systems are termed "short-wavelength" systems. "Long-wavelength" systems use InGaAsP and InP sources operating near 1.3 μm and 1.55 μm. These sources are less well developed than the $Al_xGA_{1-x}As$ sources, but power loss in an optical fiber at the longer wavelengths is lower than at 0.85 μm and the dispersion is near zero.

The fiber carries the light toward the receiver, where the light is detected and the digital signal is recovered. Since absorption, scattering and dispersion in the fiber degrade the signal, optical amplifiers are needed to regenerate the signal. Current technology

Optical Sensing and Measurement

usually requires repeaters every few kilometers.

  • An early signal repeater consisted of a detector, an amplifier, and a signal regenerator(see Figure 3.19). A conversion from optical to electrical signal, and a reconversion from electrical to optical signal was needed.

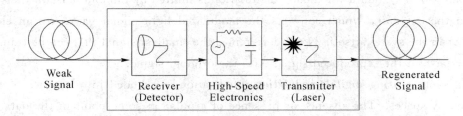

Figure 3.19 An early signal repeater

A modern repeater consists of an optical amplifier, a fiber-laser without a laser cavity (see Figure 3.20). A fiber laser consists of a properly doped fiber which can be pumped with an external light sources to produce stimulated emission. A typical fiber amplifier works in the 1550 nm band. It consists of a length of fiber doped with Erbium, which is pumped with a 980 nm laser. This pump laser supplies the energy for the amplifier, while the incoming signal stimulates emission as it passes through the doped fiber. The stimulated emission stimulates more emission. The number photons in the doped fiber increases exponentially. Gains of greater than 40 dB are possible.

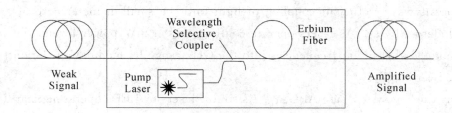

Figure 3.20 A modern repeater

The receiver converts the optical signal exiting the fiber back into an electrical signal. The receiver consists of an optical detector and a signal-conditioning circuit. The optical detector can be either a semiconductor PIN diode, whose electrical conductivity is a function of the intensity and wavelength of the light signal, or an avalanche photodiode detector.

For $Al_xGA_{1-x}As$ sources, silicon photodiodes are suitable detectors. Silicon photodiodes offer excellent high-frequency response at wavelengths up to 1.1 μm. They have peak spectral response near 0.9 μm, close to the wavelength of $Al_xGA_{1-x}As$ lasers. At longer wavelengths, in particular at 1.3 μm, germanium or InGaAsSb photodiodes must be used.

A fiber optic data link also includes passive components other than an optical fiber. Passive components used to make fiber connections affect the performance of the data link. Fiber optic components used to make the optical connections include optical splices, connectors, and couplers.

The maximum speed at which a signal may be transmitted through an optical fiber is limited by the rate at which electronics can modulate the source and by dispersion in the fiber. Currently the electronics can modulate the source at a maximum frequency of ~10 GHz. This limits the transfer rate in a single mode fiber. To transmit information at a rate of greater than 10 Gbit/s wavelength division multiplexing (WDM) is used. Multiple signals each at a different wavelength are transmitted through the same optical fiber.

The multiplexer and demultiplexer typically consist of some type of optical diffraction grating(see Figure 3.21). The total number of channels, or wavelengths ranges from 2 to over 100. The separation between wavelengths, $\Delta\lambda = (\lambda^2/c)\Delta f$, ranges from 0.4 to 3.2 nm, with the nominal wavelength about 1 550 nm.

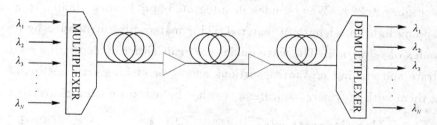

Figure 3.21 The multiplexer and demultiplexer

3.6 Integrated Optics

The goal of integrated optics (IO) is to develop miniaturized optical devices of high functionality on a common substrate. The state-of-the-art of integrated optics is still far behind its electronic counterpart. Today, only a few basic functions are commercially

Optical Sensing and Measurement

feasible. However, there exists a growing interest in the development of more and more complex integrated optical devices.

In IO we distinguish between **optical integrated circuits**, which perform functions similar to electronic circuits in communications systems, and **planar optical devices**, which are integrated optical systems other than communication systems.

Researchers hope to put wave guides, modulators, switches, and other active optical functions onto various substrates. It is visualized that thin films and micro-fabrication technologies can suitably be adopted to realize optical counterparts of integrated electronics for signal generation, modulation, switching, multiplexing and processing.

In optical integrated circuits, light is confined in thin film wave guides that are deposited on the surface or buried inside a substrate. Glasses, dielectric crystals and semiconductors can be used as substrate materials. The functions that can be realized depend on the type of substrate used. Researchers are challenged to identifying materials which have both the right electro-optical properties and a reliable means of forming them into useful structures on the integrated circuit. Unfortunately, many current-generation materials that are used to fabricate monolithic optical devices have high attenuation and therefore high transmission losses.

The most fundamental building block of an optical integrated circuits is a **channel wave guide** (see Figure 3.22). Wave guides in integrated-optics work similar to conventional fibers, trapping light in a length of material. This material is surrounded by material of a different index of refraction. The wave guides are made either by depositing material on top of a substrate and etching unwanted portions away, or etching trenches in the substrate and filling them with polymers, silicates, or other light-transmitting materials.

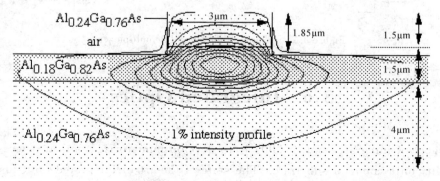

Figure 3.22 A channel wave guide

The basic requirements of a thin film optical guide material are that it be transparent to the wavelength of interest and that it have a refractive index higher than that of the medium in which it is embedded(see Figure 3.23).

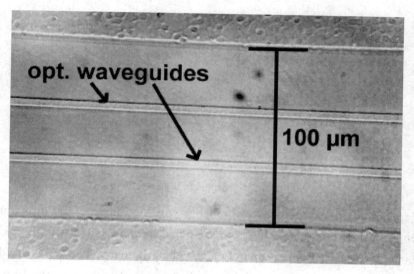

Figure 3.23   A thin film optical guide material

Within an optical integrated circuit, light propagates as a guided wave in a dielectric thin film. Appropriate addition of functional devices between interconnecting wave guides enables the realization of an optical integrated circuit for a specific use.

Some devices can be made on planar substrates using standard lithographic processes and thin-film technologies. Electron beam writing and laser beam writing are increasingly employed to produce patterns with high resolution. Epitaxial methods are used in the fabrication of sources, detectors and opto-electronic circuits on GaAs, Si and InP.

For a square channel wave guide of depth d the number of modes is approximately m^2, where $m=(2d/\lambda)$ NA, $m \gg 1$. For a slab wave guide the number of modes is approximately m. For single mode operation in any channel wave guide we therefore need $(2d/\lambda)$ NA $\leqslant$ 1, where d is the width of the channel. A single mode wave guide can be constructed by making d and NA sufficiently small. NA is the numerical aperture and for small Δ is given by

$$NA = n_{core}(2\Delta)^{1/2}, \text{ where } \Delta = (n_{core} - n_{cladding})/n_{core}$$

A **branch** in a wave guide consists of a single input wave guide, a short tapered section, and two output wave guides(see Figure 3.24). Branches are used to divide a beam

Optical Sensing and Measurement

into many channels, but the loss at a branch is always significant. Cascading n branches results in a star coupler with 2^n outputs.

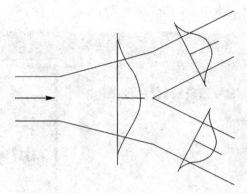

Figure 3.24 A branch in a wave guide

Directional couplers can be used as branches with less loss. Many directional couplers work by coupling the evanescent wave into the adjacent wave guide(see Figure 3.25). The wave guides must be spaced approximately a wavelength apart. The wave that is excited in the second wave guide propagates in the same direction as the wave in the first guide. The spacing between the wave guides must be adjusted for the desired power transfer.

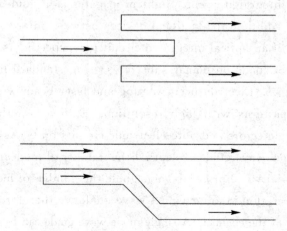

Figure 3.25 Couplers work by coupling the evanescent wave into the adjacent wave guide

Another type of coupler is based on Bragg reflection. A **fiber Bragg grating** is made from a section of ordinary single-mode optical fiber, typically a few millimeters to a few centimeters in length (see Figure 3.26). The grating is formed by causing periodic variations in the index of refraction of the glass lengthwise along the fiber. The period of the index modulation can be designed to cause deflection of light at a specific wavelength, the Bragg wavelength. Typically the light at the Bragg wavelength is selectively reflected while all other wavelengths are transmitted, essentially unperturbed by the presence of the grating.

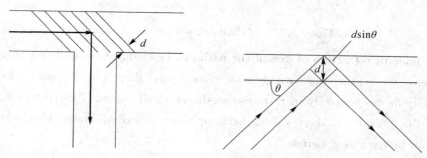

Figure 3.26 A fiber Bragg grating

The condition for Bragg reflection is $2d\sin\theta = m\lambda'$, $m = 1, 2, \ldots$, where $\lambda' = \lambda_{\text{free}}/n$ is the wavelength of the light in the fiber. By combining fiber gratings in various arrangements, many different wavelengths can be separated and coupled out. A fiber Bragg grating's low-loss characteristics and its ability to selectively pass or reflect specific frequencies, make it a very versatile element that can be used in filters and de-multiplexers for wavelength division multiplexing networks.

Bragg reflectors can also be implemented in a channel wave guide. Diffraction gratings can be etched into the wave guide. The width of the channel compared to the wavelength of the light limits the number of grooves in the grating and therefore the finesse (the effective number of transmitted beams) of the grating. The more grooves a grating has, the sharper are the diffraction peaks.

For a grating the instrumental line width $\Delta\lambda$, or the wavelength difference between two peak that can just be resolved in mth order is $\Delta\lambda = \lambda/mN$, where N is the number of grooves. If a fiber or waveguide are used to transport signals at many different wavelengths, and these signals are supposed to be separated using a grating, the wavelength difference between two signals adjacent in wavelength must be greater than $\Delta\lambda = \lambda/N$.

A wave guide that splits into two equal-length sections which after a certain distance recombine becomes a Mach-Zehnder interferometer (see Figure 3.27).

Optical Sensing and Measurement

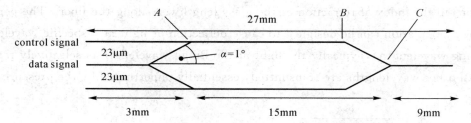

Figure 3.27 A Mach-Zehnder interferometer

If the index of refraction of one of the path can be changed by applying some external signal, then, when the the two beams recombine, they can be made to interfere constructively or destructively in response to the external signal. Destructive interference can reduce the intensity of the resulting beam or turn it off completely. The interferometer therefore can be used as a switch.

Mechanically compressing some materials using a piezoelectric effect or an acoustic wave can change their index of refraction enough to accomplish the switching function. Some materials change their index of refraction in response to heat. A microscopic resistance heater or piezoelectric transducer in one leg of an interferometer made of such a material can be used to accomplish the switching function. However, switching speeds are low for acousto-optic (tens of microseconds) and thermo-optic (milliseconds) devices.

Some materials change their optical character in response to the application of an electrical signal. The current material of choice is lithium niobate ($LiNbO_3$), which is a crystalline dielectric material with excellent electro-optical, acousto-optical and nonlinear optical properties. Mature technologies for the fabrication of optical wave guides in this material exist. The response speed of $LiNbO_3$ devices is in the nanosecond range. The drawback of $LiNbO_3$ is its high attenuation.

Examples of light-on-a-chip devices

• **Filters** are used to select desired wavelengths out of a wavelength division multiplexer (WDM) stream, or block a particular channel from being injected into a multiplexer.

Semiconductor lasers can be made into very effective filters for some applications if they operated below their threshold point. They have sharp and narrow filter characteristics, and can filter a signal as well as produce gain.

- **On/off switches** can be created using many technologies, including Mach-Zehnder-effect devices.
- **Multiplexers/demultiplexers** can be constructed for WDM applications to combine light from multiple sources, as well as extract a particular channel from a WDM beam.
- **Integrated optical amplifiers** are replacing large spools of erbium-doped fiber that were pumped with laser energy to regenerate weak network signals. The most common integrated amplifier consists of a forward-biased heterojunction which carries current. It also contains a set of optical wave guides that confine the incoming signal to the junction's active region. These semiconductor optical amplifiers can be used simply for amplification, or tuned to provide frequency conversion. This ability to shift frequencies can be exploited to build many kinds of WDM switching architectures.

The advantages of IO elements over their bulk optic counterparts are compact size, protection from thermal drift, moisture and vibration, low power requirements and low cost due to the possibility of batch fabrication.

Chapter 4   Stimulated Emission Devices Lasers

4.1   Light Amplification, Resonators

A LASER (Light Amplification by Stimulated Emission of Radiation) is an electromagnetic oscillator which combines light amplification with feedback. The laser uses mirrors to feed the light output from an optical amplifier through a delay (the travel time of the light) back into the amplifier input.

A laser is capable of producing an intense beam of photons having identical scalar and vector properties (frequency, phase, direction and polarization). As a result, a laser beam can be bright, monochromatic, coherent, and unidirectional.

Quantum mechanics predicts that all atomic and molecular systems are characterized by discreet energy levels. These energy levels are the eigenvalues of the Hamiltonian of the system and the corresponding states are its eigenstates. The state corresponding to the lowest possible energy level is termed the ground state and the states corresponding to the other levels are termed excited states. If the energy of the system is measured at any time, the result of the measurement will be one of the energy eigenvalues, and after the measurement the system will be in the corresponding eigenstate.

In a dense medium, such as a solid, liquid, or high pressure gas, frequent collisions between the atoms or molecules cause transitions between energy levels. Optically allowed transitions can also change the energy of the system. An optically allowed transition between energy levels involves the absorption or emission of a photon with frequency f, such that $hf = \Delta E$, where ΔE is the energy difference between the initial and final energy level and h is Planck's constant. If the angular momentum quantum number l_i of the initial state and the angular momentum quantum number l_f of the final state differ by l, i.e. if $\Delta l = \pm 1$, then a transition involving a photon is very likely. If $\Delta l \neq \pm 1$, then a transition involving a photon is much less likely. If no lower lying state satisfying $\Delta l = \pm 1$ exists, an excited state is called meta-stable. $\Delta l = \pm 1$ is called a **selection rule**. If this selection rule is

Chapter 4 Stimulated Emission Devices Lasers

not satisfied, an optical transition is forbidden (unlikely).

An atom or molecule can emit a photon via spontaneous emission or stimulated emission (see Figure 4.1). Atoms and molecules in excited states randomly emit single photons in all directions according to statistical rules via spontaneous emission. In the process of stimulated emission, a photon of energy hf perturbs an excited atom or molecule and causes it to relax to a lower level, emitting a photon of the same frequency, phase, and polarization as the perturbing photon. Stimulated emission is the basis for photon amplification and it is the fundamental mechanism underlying all laser action. The quantum mechanical treatment of stimulated emission is very similar to that of absorption.

Consider the simple case of a two-level system with lower level 1 and upper level 2. Let N_1 be the number density of atoms in level 1 and N_2 be the number density in level 2. Let $u(f_{12})$ be the energy density per unit frequency interval of the light at frequency $f_{12} = (E_2 - E_1)/h$.

- The rate of spontaneous emission is independent of $u(f_{12})$.

It is proportional to N_2.

$R_{\text{spon. emiss.}} = A_{21} N_2$.

- The rate of stimulated emission depends on $u(f_{12})$.

It is proportional to $u(f_{12}) N_2$.

$R_{\text{stim. emiss.}} = B_{21} u(f_{12}) N_2$.

- The rate absorption depends on $u(f_{12})$.

It is proportional to $u(f_{12}) N_1$.

$R_{\text{absorb.}} = B_{12} u(f_{12}) N_1$.

The proportional constants A_{21}, B_{21}, and B_{12} are called the **Einstein coefficients**.

Simple quantum mechanics predicts $B_{21} = B_{12}$ and lets us calculate the value of $B_{21} = B_{12}$ using time-dependent perturbation theory. But as long as we treat the electromagnetic field classically, we cannot calculate the probability for spontaneous emission of a photon this way. To calculate A_{21} we also need to quantize the radiation field.

We can however avoid this problem by making statistical arguments. In a cavity in thermal equilibrium the probabilities that states 1 and 2 are occupied are proportional to the Boltzmann factors $\exp(-E_1/(kT))$ and $\exp(-E_2/(kT))$, respectively, and in **equilibrium** the probability of up transitions must exactly balance the probability of down transitions. We have

$$N_1 \propto \exp\left(\frac{-E_1}{kT}\right), \quad N_2 \propto \exp\left(\frac{-E_2}{kT}\right)$$

We therefore need

Optical Sensing and Measurement

$$(A_{21}+B_{21}u(f_{12}))\exp\left(\frac{-E_2}{kT}\right)=B_{12}u(f_{12})\exp\left(\frac{-E_1}{kT}\right)$$

$$=B_{21}u(f_{12})\exp\left(\frac{-E_1}{kT}\right) \tag{4.1}$$

$$(A_{21}+B_{21}u(f_{12}))=B_{21}u(f_{12})\exp\left(\frac{E_2-E_1}{kT}\right)=B_{21}u(f_{12})\exp\left(\frac{hf_{12}}{kT}\right) \tag{4.2}$$

$$A_{21}=B_{21}u(f_{12})\left[\exp\left(\frac{hf_{12}}{kT}\right)-1\right] \tag{4.3}$$

In a cavity in thermal equilibrium $u(f_{12})$ is given by Plank's law,

$$u(f_{12})=\frac{8\pi hf^3/c^3}{\exp(\frac{hf_{12}}{kT})-1} \tag{4.4}$$

We therefore have

$$A_{21}=\frac{B_{21}8\pi hf^3}{c^3}.$$

The competition between absorption, stimulated emission, and spontaneous emission defines the criteria for laser action.

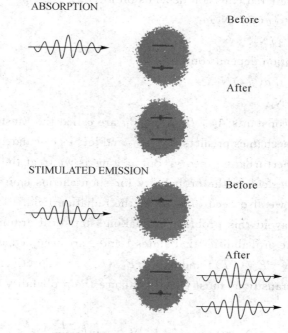

Figure 4.1 The stimulated emission

In thermal equilibrium $N_1>N_2$, a resonant photon is more likely to be absorbed than to stimulate emission. But if $N_2 > N_1$ there is the possibility of average overall amplification for an array of photons passing through a volume of atoms of the two-level system. This situation is termed **population inversion**, since under normal conditions $N_1 > N_2$.

Spontaneous emission depletes N_2 at a rate proportional to A_{21}, producing unwanted photons with random phases, propagation directions, and polarizations. Because of loss associated with spontaneous emission and other losses associated with the laser cavity, each laser is characterized by a minimum value of $N_2 - N_1$, termed the **threshold inversion**. Only if $N_2 - N_1$ is greater then the threshold inversion do we see laser action.

In a simple helium neon laser(see Figure 4.2), Two sets of excited states of neon ($2s$ and $3s$) occur at excitation energies similar to those of the 2^3S and 2^1S states of helium, which are favorably populated by the laser discharge. Within the plasma, atoms are rapidly colliding with other atoms, electrons, and with the walls of the tube. Collisions between excited state helium atoms and ground state neon atoms cause some of the neon atoms to be excited to the $2s$ and $3s$ states. Collisional energy transfer is a resonant process in which the

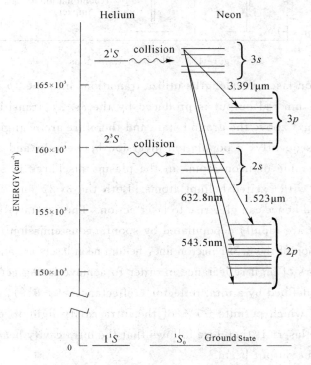

Figure 4.2　States of neon and helium

total energy is conserved. The small energy mismatch is compensated by changes in the kinetic energy of the atoms (see Figure 4.3).

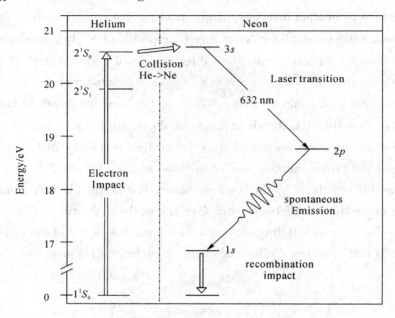

Figure 4.3 The laser transitions

Most of the neon laser wavelengths utilize transitions into the $2p$ series of levels. For example, the 632.8 nm red output is produced by the $3s_2$-$2p_4$ transition. The $2p$ energy levels lie 150 000 cm^{-1} above the ground state and therefore are negligibly populated, even in the plasma. Consequently, a population inversion between $3s$ and $2p$, and between $2s$ and $2p$, in the ensemble of neon atoms in the plasma discharge is easily maintained by collisional pumping with excited helium atoms. Both the $3s$-$2p$ and $2s$-$2p$ transitions are optically allowed and hence can give rise to laser action. The $2p$ states are populated by the laser transitions but are rapidly depopulated by spontaneous emission to lower levels.

With the exception of the 3.391 micron line, helium neon lasers are all low gain devices and require cavity mirrors of high reflectance in order to achieve lasing action. Typically, one end of the cavity is defined by a total reflector (reflectance$>$99.9%), and the other end is defined by a mirror which permits $\sim$1% of the intra cavity light to escape, forming the useful output of the laser. It therefore follows that the intra cavity beam is up to 200 times more intense than the output beam.

The $3s_2$-$2p_{10}$ transition in neon at 543.5 nm is a very low gain transition. To make a

Chapter 4 Stimulated Emission Devices Lasers

green He-Ne laser this transition is made to lase in a compact plasma tube by optimizing the tube design, gas mixture, processing techniques, and cavity mirror specifications. Since prisms are not desirable in commercial helium neon lasers (for economy and stability), cavity mirrors have to meet stringent wavelength dependent specifications in order to cause amplification at 543.5 nm, while suppressing the 632.8 nm line and the very high gain 3.39 micron line.

Various combinations of cavity mirror types and separations have been used with helium neon lasers(see Figure 4.4).

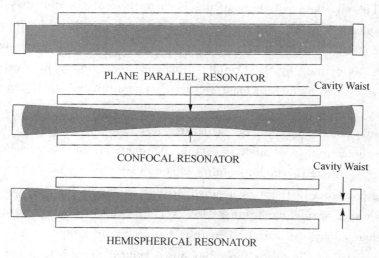

Figure 4.4 The resonator

- The simplest configuration of two plane parallel mirrors is almost never used because it produces a cavity with high diffraction losses and is very sensitive to misalignment.
- The confocal resonator uses two concave, spherical mirrors with equal radii. Each is placed at the center of curvature of the other. This cavity utilizes a large fraction of the plasma volume and produces higher power. It is much less sensitive to misalignment, but it is sensitive to exact mirror separation. The confocal resonator is rarely used when TEM_{00} output is required.
- The hemispherical resonator has a plane mirror at the center of curvature of a concave spherical mirror. This configuration is easy to adjust and produces highly coherent output. Such cavities are the most commonly used in commercial He-Ne lasers as they yield

an excellent combination of good power, ease of adjustment, stability once aligned and good mode control.

A gas laser transition does not produce an infinitesimally narrow line. Each line has a natural line width and the Doppler effect and pressure effects broadened it even more. If the Doppler effect and pressure effects dominate the line width, the it will have a smooth Gaussian profile. Red He-Ne lasers typically have a full-width-at-half-maximum (FWHM) of 1400 MHz. Superimposed on the Doppler broadened gain curve is the cavity resonance function, with the mode spacing given by $f = c/2L$. For an 0.5 m cavity, the mode spacing is 300 MHz. The sharpness of these modes is a function of the multipass nature of the cavity and is typically 1 MHz. For such a cavity, laser output therefore consists of five or six single sharp lines (FWHM 1 MHz) separated by 300 MHz. The relative intensity of the cavity modes is defined by the Doppler broadened gain curve. In a very short cavity ($L < 0.15$ m), only one mode will exist and the laser output will then consist of a single frequency.

The multiple longitudinal mode structure described above gives rise to a power fluctuation phenomenon termed mode sweeping. All unstabilized helium neon lasers exhibit this effect which is due to thermal instability causing variation in the cavity length, ΔL. As the cavity length changes, there is a small change in mode spacing which is typically 10 kHz or less under normal conditions. However, the absolute wavelength of each cavity mode is also changed by the variation in tube length by an amount $\Delta \lambda = (\Delta L/L)\lambda$. $\Delta \lambda$ depends on the glass type used for the tube, but is typically $\sim 2.5 \times 10^{-3}$ nm/℃. In effect, the "comb" of longitudinal modes drifts with respect to the Doppler broadened line center, repeating its initial relative position in less than 1℃. Because of the Gaussian profile of the gain curve, changes in the mode "comb" position with respect to line center can cause changes in the overall power output. If the mode spacing is very small, as with a long laser tube, these changes may be very small. On the other hand, a short laser tube may have only one or two cavity modes under the Doppler profile, and the sum of their intensity is very dependent on their position on the Gaussian gain curve.

This effect is almost identical for all unstabilized commercial TEM_{00} tubes and is a function of cavity length. The overall amplitude fluctuations are typically a few percent.

In a simple helium neon laser with cylindrical symmetry, the output is said to be randomly polarized. Each individual cavity mode has a linear polarization at any one time. However, the overall laser output is a time varying mix of modes of different polarization.

Consequently, the beam appears to be non-polarized when integrated over a fairly short period of time.

Although the beam intensity is fairly constant, if the experiment or application involves polarization dependent optics, such as beam-splitters, then large rapid amplitude fluctuations will be apparent. If this is not acceptable, a polarized helium neon laser must be used. A polarized helium neon laser has an intra cavity Brewster window which introduces sufficient loss in the s-plane of polarization (defined by the window) so that only p-polarized output is produced.

There is an interesting and sometimes useful polarization relationship between adjacent cavity modes in a randomly polarized helium neon laser. The He-Ne is a low gain amplifier with only a limited number of excited neon atoms in the 2 and 3 states. The gain saturates at a low level. As a free-run amplifier, the laser tends to oscillate only on the modes of highest gain, a process known as gain clamping. A complete quantum mechanical treatment of the problem shows that saturation occurs at higher overall levels if adjacent cavity modes are polarized perpendicular to each other. This effect is indeed observed in randomly polarized lasers.

The amplification bandwidth of a laser is usually broader than the cavity mode spacing. This width is associated with the actual transition width and broadening effects such as the Doppler effect. As a consequence, many lasers simultaneously run on several cavity modes. In lasers with radial symmetry, the polarization of a particular longitudinal mode is arbitrary and changes with time unless controlled. However, in low gain devices such as the helium neon laser, amplification is enhanced when adjacent modes are polarized perpendicular to each other. This reduces the number of atoms excluded from the stimulated emission process. Since the phenomenon of gain-clamping causes a laser to operate naturally at its highest gain, the adjacent modes of such a laser are polarized perpendicular to each other unless otherwise regulated.

The gain per pass varies greatly between different types of lasers and so do mirror requirements. Pulsed lasers tend to exhibit higher gain per pass than continuous wave (CW) lasers, although the massive population inversion can only be maintained for a small interval of time. A typical helium-neon laser has a single pass gain of the order of 1% and requires high reflectance mirrors (100% and 99%). Lasers that are capable of producing multiple wavelengths from several different lasing transitions require specially optimized mirrors if output at each wavelength is desired.

Optical Sensing and Measurement

In addition to the longitudinal mode structure, the radiation field may have nodes and antinodes in the plane perpendicular to the laser axis (see Figure 4.5). The different transverse irradiance patterns are referred to as Transverse Electric and Magnetic (TEM) modes. In an unconfined cavity with cylindrical symmetry, these modes form simple progressions of irradiance functions. Transverse modes are identified by their irradiance distributions and are designated TEM_{pq} where the subscripts p and q refer to the number of nodes along the two orthogonal axes perpendicular to the laser axis. The irradiance distributions of all the modes are smoothly varying but none has a uniform irradiance distribution. The lowest order mode, TEM_{00} has a cylindrical Gaussian irradiance distribution (see Figure 4.6).

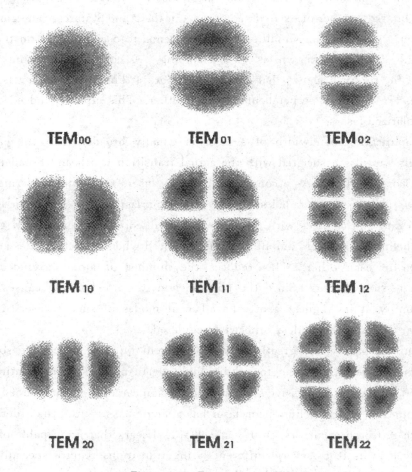

Figure 4.5 Transverse modes

Chapter 4 Stimulated Emission Devices Lasers

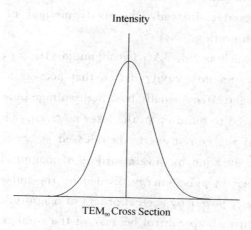

Figure 4.6 TEM$_{00}$ mode

This mode experiences the minimum possible diffraction loss, has minimum divergence, and can be focused to the smallest possible spot. For these reasons, it is often imperative that the laser be restricted to operation in this mode. Higher order modes have a larger spread and suffer higher diffraction losses.

Some lasers use an internal tuning element to select various lines. In an argon-ion laser, which has strong output at 488 and 514 nm as well as several other wavelengths, the operational wavelength is selected by a prism within the cavity that is rotated to an angle that provides single-line output.

Various liquid and solid-state lasers have broad bandwidths that cover tens of nanometers. Examples include dye and Ti:sapphire lasers. This has allowed the development of tunable and ultra fast lasers. Creating a tunable CW laser involves including an extra filtering element in the cavity, usually a birefringent filter. This narrows the bandwidth and, by rotating the filter, allows smooth bandwidth tuning.

Lasers can be divided into three main categories, continuous wave (CW), pulsed, and ultra fast. Some materials such as ruby and rare-gas halogen excimers (ArF, XeCl) sustain laser action for only a brief period. If the pulse duration is sufficiently long (microseconds), laser design is similar to that of a CW laser. However, many pulsed lasers are designed for pulse duration of a few nanoseconds. The light in each pulse cannot make many round-trips in the cavity. Resonant cavity designs used in CW lasers cannot control such a laser. The pulse dies before equilibrium conditions are reached. So, while two mirrors are still used in pulsed lasers for defining the direction of highest gain, they do not act as a resonant cavity. Instead, the usual method of controlling and tuning wavelength is a diffraction grating.

Some pulsed lasers, such as Nd:YAG (neodymium yttrium aluminum garnet) can be operated with a **Q-switch**, an intra-cavity device that acts as a fast optical gate. Light cannot pass it unless it is activated, usually by a high-voltage pulse. Initially, the switch is closed and energy is allowed to build up in the laser material. Then at the optimum time, the switch is opened and the stored energy is released as a very short pulse. This can shorten the normal pulse duration by several orders of magnitude. The peak power of a pulsed laser is proportional to pulse energy divided by the pulse duration. Q-switching, therefore increases the peak power by several orders of magnitude. The wavelength purity of Q-switched lasers is difficult to control because of the combination of high peak power and short pulse duration. To solve this problem, a low-power, well-controlled oscillator is often put in series with one or more amplifiers.

CW lasers can produce many longitudinal modes. If the cavity is pulsed or oscillated, it is possible to lock these modes together. The resultant interference causes the traveling light waves inside the cavity to collapse into a very short pulse. (We build a wave packet.) Every time this pulse reaches the output coupler, the laser emits a part of this pulse. The pulse repetition rate is determined by the time it takes for the pulse to make one trip around the cavity. The more modes interfere, the shorter is the pulse duration. The pulse duration is inversely proportional to the bandwidth of the laser gain material. The materials commonly used for tunable lasers produce the shortest mode-locked pulses. Popular materials include Ti:sapphire and similar materials. Turnkey commercial Ti:sapphire lasers now deliver pulses as short as 20 fs, with typical repetition rates around 100 MHz and peak powers approaching 1 MW.

For many years the most common CW laser was the He-Ne laser. These low-power (mW) lasers use an electric discharge to create the population inversion. Most He-Ne laser emit in the red at 633 nm. They are used in interferometers, bar-code reader, and laser printers. They also are used for many sighting and pointing applications from medical surgery to high-energy physics.

In a helium neon laser, a low pressure mixture of helium and neon ($<15\%$) is contained in a narrow bore glass tube. A longitudinal, DC electrical discharge is maintained in the narrow tube. This creates a plasma. Collisions with free electrons in the plasma excite the helium atoms and cause a significant number of them to be trapped in the lowest energy metastable states. These are the 2^1s and the 2^3s states, where one of the two helium electrons has been raised from the lowest energy 1s atomic orbital to the 2s orbital. For this reason, they are termed electronically excited states. Neon is a larger, more complex atom, with 10 electrons which are arranged in a $1s^2 2s^2 2p^6$ configuration in the $^1S_0$ ground state. Neon has many excited states, and the ones concerned with laser action are shown in the energy level diagram below. The multiplet nature of the electronically excited states arises from the number of different ways in which the angular momenta of the electrons can be combined.

4.2 Types and Operating Principles

Gas lasers

Gas lasers using many gases have been built and used for many purposes(see Figure 4.7).

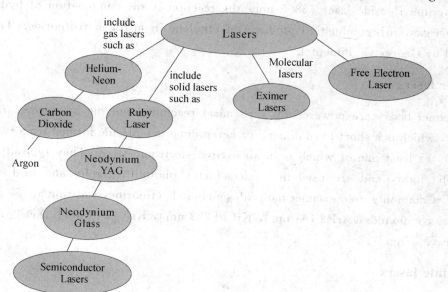

Figure 4.7 Laser types

The helium-neon laser (He-Ne) emits at a variety of wavelengths and units operating at 633 nm are very common in education because of its low cost.

Carbon dioxide lasers can emit hundreds of kilowatts at 9.6 μm and 10.6 μm, and are often used in industry for cutting and welding. The efficiency of a CO_2 laser is over 10%.

Argon-ion lasers emit light in the range 351 – 528.7 nm. Depending on the optics and the laser tube a different number of lines is usable but the most commonly used lines are 458 nm, 488 nm and 514.5 nm.

A nitrogen transverse electrical discharge in gas at atmospheric pressure (TEA) laser is an inexpensive gas laser producing UV Light at 337.1 nm.

Metal ion lasers are gas lasers that generate deep ultraviolet wavelengths. Helium-silver (HeAg) 224 nm and neon-copper (NeCu) 248 nm are two examples. These lasers have particularly narrow oscillation linewidths of less than 3 GHz (0.5 picometers), making them candidates for use in fluorescence suppressed Raman spectroscopy.

Chemical lasers

Chemical lasers are powered by a chemical reaction, and can achieve high powers in continuous operation. For example, in the Hydrogen fluoride laser (2700 – 2900 nm) and the Deuterium fluoride laser (3800 nm) the reaction is the combination of hydrogen or deuterium gas with combustion products of ethylene in nitrogen trifluoride. They were invented by George C. Pimentel.

Excimer lasers

Excimer lasers are powered by a chemical reaction involving an *excited dimer*, or *excimer*, which is a short-lived dimeric or heterodimeric molecule formed from two species (atoms), at least one of which is in an excited electronic state. They typically produce ultraviolet light, and are used in semiconductor photolithography and in LASIK eye surgery. Commonly used excimer molecules include F_2 (fluorine, emitting at 157 nm), and noble gas compounds (ArF [193 nm], KrCl [222 nm], KrF [248 nm], XeCl [308 nm], and XeF [351 nm]).

Solid-state lasers

Solid state laser materials are commonly made by doping a crystalline solid host with ions that provide the required energy states. For example, the first working laser was a ruby laser, made from ruby (chromium-doped corundum). Formally, the class of solid-state lasers includes also fiber laser, as the active medium (fiber) is in the solid state. Practically, in the scientific literature, solid-state laser usually means a laser with bulk active medium; while wave-guide lasers are caller fiber lasers.

Neodymium is a common dopant in various solid state laser crystals, including yttrium orthovanadate (Nd:YVO$_4$), yttrium lithium fluoride (Nd:YLF) and yttrium aluminium garnet (Nd:YAG)(see Figure 4.8). All these lasers can produce high powers in the infrared spectrum at 1064 nm. They are used for cutting, welding and marking of metals and other materials, and also in spectroscopy and for pumping dye lasers. These lasers are also commonly frequency doubled, tripled or quadrupled to produce 532 nm (green, visible), 355 nm (UV) and 266 nm (UV) light when those wavelengths are needed.

Figure 4.8　A 50 W FASOR, based on a Nd:YAG laser, used at the Starfire Optical Range

Ytterbium, holmium, thulium, and erbium are other common dopants in solid state lasers. Ytterbium is used in crystals such as Yb:YAG, Yb:KGW, Yb:KYW, Yb:SYS, Yb:BOYS, Yb:CaF2, typically operating around 1020-1050 nm. They are potentially very efficient and high powered due to a small quantum defect. Extremely high powers in ultrashort pulses can be achieved with Yb:YAG. Holmium-doped YAG crystals emit at 2097 nm and form an efficient laser operating at infrared wavelengths strongly absorbed by water-bearing tissues. The Ho-YAG is usually operated in a pulsed mode, and passed through optical fiber surgical devices to resurface joints, remove rot from teeth, vaporize cancers, and pulverize kidney and gall stones.

Titanium-doped sapphire (Ti:sapphire) produces a highly tunable infrared laser, commonly used for spectroscopy as well as the most common ultrashort pulse laser.

Thermal limitations in solid-state lasers arise from unconverted pump power that manifests itself as heat and phonon energy. This heat, when coupled with a high thermo-optic coefficient (dn/dT) can give rise to thermal lensing as well as reduced quantum efficiency. These types of issues can be overcome by another novel diode-pumped solid state laser, the diode-pumped thin disk laser. The thermal limitations in this laser type are mitigated by utilizing a laser medium geometry in which the thickness is much smaller than the diameter of the pump beam. This allows for a more even thermal gradient in the material. Thin disk lasers have been shown to produce up to kilowatt levels of power.

Fiber-hosted lasers

Solid-state lasers where the light is guided due to the total internal reflection in an optical fiber are called fiber lasers. Guiding of light allows extremely long gain regions providing good cooling conditions; fibers have high surface area to volume ratio which allows efficient cooling. In addition, the fiber's waveguiding properties tend to reduce thermal distortion of the beam. Erbium and ytterbium ions are common active species in such lasers.

Quite often, the fiber laser is designed as a double-clad fiber. This type of fiber consists of a fiber core, an inner cladding and an outer cladding. The index of the three concentric layers is chosen so that the fiber core acts as a single-mode fiber for the laser emission while the outer cladding acts as a highly multimode core for the pump laser. This lets the pump propagate a large amount of power into and through the active inner core region, while still having a high numerical aperture (NA) to have easy launching conditions.

Pump light can be used more efficiently by creating a fiber disk laser, or a stack of such lasers.

Fiber lasers have a fundamental limit in that the intensity of the light in the fiber cannot be so high that optical nonlinearities induced by the local electric field strength can become dominant and prevent laser operation and/or lead to the material destruction of the fiber. This effect is calledphotodarkening. In bulk laser materials, the cooling is not so efficient, and it is difficult to separate the effects of photodarkening from the thermal effects, but the experiments in fibers show that the photodarkening can be attributed to the formation of long-living color centers.

Photonic crystal lasers

Photonic crystal lasers are lasers based on nano-structures that provide the mode confinement and the density of optical states (DOS) structure required for the feedback to take place. They are typical micron-sized and tunable on the bands of the photonic crystals.

Semiconductor lasers

Commercial laser diodes emit at wavelengths from 375 nm to 1800 nm, and wavelengths of over 3 μm have been demonstrated(see Figure 4.9). Low power laser diodes are used in laser printers and CD/DVD players. More powerful laser diodes are frequently used to optically pump other lasers with high efficiency. The highest power industrial laser diodes, with power up to 10 kW (70 dBm), are used in industry for cutting and welding. External-cavity semiconductor lasers have a semiconductor active medium in a larger cavity. These devices can generate high power outputs with good beam quality, wavelength-tunable narrow-linewidth radiation, or ultrashort laser pulses.

Figure 4.9 A typical commercial laser diode, probably from a CD or DVD player

Vertical cavity surface-emitting lasers (VCSELs) are semiconductor lasers whose emission direction is perpendicular to the surface of the wafer. VCSEL devices typically have a more circular output beam than conventional laser diodes, and potentially could be much cheaper to manufacture. As of 2005, only 850 nm VCSELs are widely available, with 1300 nm VCSELs beginning to be commercialized, and 1550 nm devices an area of research. VECSELs are external-cavity VCSELs. Quantum cascade lasers are semiconductor lasers that have an active transition between energy *sub-bands* of an electron in a structure containing several quantum wells.

The development of a silicon laser is important in the field of optical computing, since it means that if silicon, the chief ingredient of computer chips, were able to produce lasers, it would allow the light to be manipulated like electrons are in normal integrated circuits. Thus, photons would replace electrons in the circuits, which dramatically increases the speed of the computer. Unfortunately, silicon is a difficult lasing material to deal with, since it has certain properties which block lasing. However, recently teams have produced silicon lasers through methods such as fabricating the lasing material from silicon and other semiconductor materials, such as indium(Ⅲ) phosphide or gallium(Ⅲ) arsenide, materials which allow coherent light to be produced from silicon. These are called hybrid silicon laser. Another type is a Raman laser, which takes advantage of Raman scattering to produce a laser from materials such as silicon.

Dye lasers

Dye lasers use an organic dye as the gain medium. The wide gain spectrum of available dyes allows these lasers to be highly tunable, or to produce very short-duration pulses (on the order of a few femtoseconds).

Free electron lasers

Free electron lasers, or FELs, generate coherent, high power radiation, that is widely tunable, currently ranging in wavelength from microwaves, through terahertz radiation and infrared, to the visible spectrum, to soft X-ray lasers. They have the widest frequency range of any laser type. While FEL beams share the same optical traits as other lasers, such as coherent radiation, FEL operation is quite different. Unlike gas, liquid, or solid-state lasers, which rely on bound atomic or molecular states, FELs use a relativistic electron beam as the lasing medium, hence the term *free electron*.

Chapter 4 Stimulated Emission Devices Lasers

Exotic laser media

In September 2007, the BBC News reported that there was speculation about the possibility of using positronium annihilation to drive a very powerful gamma ray laser. Dr. David Cassidy of the University of California, Riverside proposed that a single such laser could be used to ignite a nuclear fusion reaction, replacing the hundreds of lasers used in typical inertial confinement fusion experiments.

Space-based X-ray lasers pumped by a nuclear explosion have also been proposed as antimissile weapons. Such devices would be one-shot weapons.

Since the early period of laser history, laser research has produced a variety of improved and specialized laser types, optimized for different performance goals, including:

- ▶ new wavelength bands
- ▶ maximum average output power
- ▶ maximum peak output power
- ▶ minimum output pulse duration
- ▶ maximum power efficiency
- ▶ maximum charging
- ▶ maximum firing

and this research continues to this day.

Optical Sensing and Measurement

Chapter 5   Sources and Detectors

5.1   The Source of Light

An important part of the design of an optical system is its efficiency in transferring light. One must be able to specify the amount of energy emitted or received. For historical reasons, many similar quantities are used to specify the amount of light leaving the source or arriving at the receiver, and many different systems of units are used. Only in recent years has the situation improved with the gradual changeover to the International System of Units (SI).

To specify the amount of energy emitted by a source, we use the quantities.
- Energy
- Power
- Power per unit area
- Power per unit solid angle
- Power per unit solid angle per unit projected area

To specify the amount of energy received by a detector, we use one quantity.
- Power per unit area

A radian is the angle subtended at the center of a circle of radius r by a section of its circumference of length equal to r. Dividing $2\pi r$ by r gives 2π as the number of radians in a full circle.

A steradian is the solid angle subtended at the center of a sphere of radius r by a section of its surface area of magnitude equal to r^2. Since the surface area is $4\pi r^2$, there are 4π steradians surrounding a point in space.

Let a cone of arbitrary shape have its apex at the center of a sphere of unit radius. The solid angle at the apex is numerically equal to the surface area on the sphere intercepted by the cone, since the full sphere has an area of 4π.

When specifying energy, power, power per unit area, power per unit solid angle, and power per unit solid angle per unit projected area, we can specify radiometric quantities, which apply to radiation of any wavelength, or photometric quantities, which only apply to visible light.

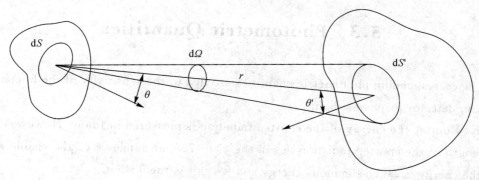

Figure 5.1 The area dS emits radiation toward a receiving area dS' at a distance r

In the figure above, let the area dS be the source of electromagnetic radiation. The area dS emits radiation toward a receiving area dS' at a distance r (see Figure 5.1).

5.2 Radiometric Quantities

- In SI units, the energy E of the emitted radiation is measured in Joule (J).
- The radiometric power $P = dE/dt = \varphi$ emitted from a surface dS is called the **radiant flux**, and it is measured in J/s = Watt (W).
- The power per unit area $dP/dS = M$ is called the **radiant emittance**. It is measured in W/m^2.
- The power per unit solid angle $dP/d\Omega = I$ is called the **radiant intensity**. It is measured in unit of W/sr. I can have an angular dependence. We then specify $I(\theta)$.
- The power per unit solid angle per unit projected area is called the **radiance** L. It is measured in $W/sr/m^2$. An important part of the definition of L is the specification of the area. We use the normal or projected area $dS \cos\theta$, since the energy received at dS' depends not only on the size of dS but also on its orientation. (The area is projected onto a plane whose normal is the line of sight.)

Optical Sensing and Measurement

$$L = \frac{I}{dS \cos\theta} = \frac{d^2 P}{d\Omega\, dS \cos\theta}$$

- The amount of energy received from a source by the surface dS' is the **irradiance** E, measured in W/m^2.

5.3 Photometric Quantities

The corresponding photometric quantities are tied to the sensitivity of the human eye, our main detector for visible light.

In SI units, the energy of the emitted radiation is measured in Joule. However, if the wavelength of the emitted radiation lies in the 400~700 nm range, i.e. the visible range, then the energy is called luminous energy and its unit is the Talbot.

- The photometric equivalent of the radiant flux φ is the **luminous flux** φ, whose units are lumens (lm).

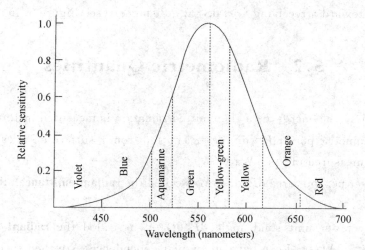

Figure 5.2 The sensitivity of the eye as a function of wavelength

The figure above shows the sensitivity of the eye as a function of wavelength or color (see Figure 5.2). The peak is in the green at 555 nm. 1 W of green light is equivalent to 683 lm.

- The photometric equivalent of the radiant emittance M is the **luminous emittance** M, measured in lm/m² (see Figure 5.3).

• The photometric equivalent of the radiant intensity I is the **luminous intensity** I. It is measured in lm/sr or candela. Candela was once known as the candle. The luminous intensity I was called the "candlepower".

• The photometric equivalent of the radiance L is the **luminance** L, measured in candela/m² or lm/sr/m². The luminance L is the property of a source that is commonly called "**brightness**". The unit for L used to be the Lambert.

• The photometric equivalent of the irradiance E is the **illuminance** E, measured in lm/m² = lux (lx).

If the energy supplied by dS is being received from another source and retransmitted isotropically, i.e. if dS is a perfect diffuser and its radiance or brightness is not a function of angle, then $L = I/(dS \cos\theta) = I/dS_n$ = constant and $I(\theta) = L dS \cos\theta$ is related to its normal component I_n by the equation

$$I(\theta) = I_n \cos\theta. \tag{5.1}$$

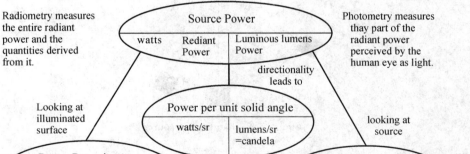

Figure 5.3 Photometric quantities

This is known as **Lambert's law**. A surface which satisfies Lambert's law is called **Lambertian**. Lambertian refers to a flat radiating surface. (The flat surface can be an elemental area of a curved surface.) A Lambertian surface can be an active surface or a passive, reflective surface. The intensity $I(\theta)$ falls off as the cosine of the observation angle with respect to the surface normal (Lambert's law). The radiance (W/(m² sr)) is independent of direction. A good example is a surface painted with a good "matte" or "flat" white paint. If it is uniformly illuminated, like from the sun, it appears equally bright from whatever direction you view it.

In the figure 5.4, if the source is Lambertian equally bright in all directions, then the amount of energy that falls each the detectors is not the same for each of the detectors, but depends on the angle θ as $\cos\theta$. [Note: A source that is equally bright in all direction does not emit the same amount of energy in all directions. When determining the brightness of a small source we measure the energy that falls on a detector subtending a given solid angle at the source and then divide by the apparent size (area) of the source. A small source emitting the same amount of energy into the given solid angle as a larger source is brighter than the larger source.]

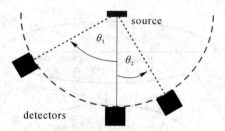

Figure 5.4 Lambertian source

The apparent brightness of a Lambertian surface is the same when viewed from any angle. For a Lambertian surface only, we have the relation ship between the emittance M and the radiance L,

$$M = \pi L$$

$[L = d^2 P/(d\Omega\, dS\, \cos\theta).$ If L is constant, then

$$M = dP/dS = \int_0^{\pi/2} L \cos\theta d\Omega = 2\pi L \int_0^{\pi/2} \cos\theta \sin\theta d\theta = \pi L.]$$

The ratio of the radiant emittance (W/m²) to the radiance (W/m² sr) of a Lambertian surface is a factor of π and not 2π. The radiance is integrated over a hemisphere.

The factor of $\cos(\theta)$ in the definition of radiance is responsible for this result. It is counterintuitive, since there are 2π steradians in a hemisphere.

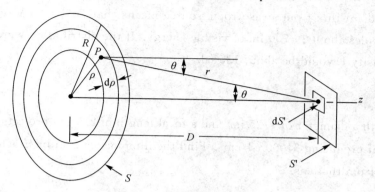

Figure 5.5 A radiating disk source and a receiving surface

The figure above (see Figure 5.5) shows a radiating disk source and a receiving surface dS' normal to the z-axis. We wish to find the connection between the luminance L of the radiating disk and the illuminance E at the receiving surface. The flux in the direction shown from an elementary area on the disk is

$$d^2\varphi = L d\Omega dS \cos\theta$$

The solid angle subtended by dS' at the radiating point P is

$$d\Omega = \frac{dS' \cos\theta}{r^2} \tag{5.2}$$

The radiation comes from the element $dS = 2\pi\rho d\rho$.
If we assume that L is constant across the disk, then

$$\varphi = 2\pi L \int dS' \int_0^R \frac{\cos^2\theta}{r^2} \rho d\rho = 2\pi L S' \int_0^R \frac{\cos^2\theta}{\rho^2 + D^2} \rho d\rho \tag{5.3}$$

Substituting for $\cos\theta$ and integrating yields

$$\varphi = \pi L S' \frac{R^2}{R^2 + D^2} \tag{5.4}$$

From the definition of the illuminance $E = d\varphi/dS'$ we obtain

$$E = \frac{\pi L R^2}{R^2 + D^2} = \frac{LS}{r^2} \tag{5.5}$$

where S is the area of the disk.
When dS' is far enough away, we may finally write

$$E = \Omega L \tag{5.6}$$

which is the desired relation.

A source with a uniform flux of 4 lm radiates 1 lm/sr or 1 candela. Hence, 4π or 12.57 lm of radiant flux from an isotropic source means that $I=1$ cd. A typical 100-watt light bulb provides about 1,700 lm of visible energy. If the distribution were uniform, the luminous intensity I would be about 135 cd.

Problem:

A lens with a diameter of 1.25 in. and a focal length of 2.0 in. projects the image of a lamp capable of producing 3,000 cd/cm^2. Find the illuminance E in lm/ft^2 (footcandles) on a screen 20 ft from the lens.

[In the figure 5.6, the area of the spherical cap on a sphere with r is $\int_0^\theta 2\pi r^2 \sin\theta' \, d\theta' = 2\pi r^2 (1-\cos\theta)$, the solid angle Ω subtended by the cap is $2\pi(1-\cos\theta)$.

For the lens $\cos\theta = X_0/(X_0^2 + R^2)^{1/2}$.]

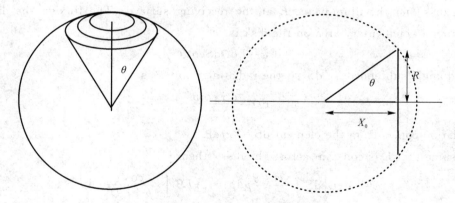

Figure 5.6 The area of the spherical cap on a sphere

The radiant energy emitted per cm^2 into a solid angle 0.282 sr is 3000 times 0.282=845.6 lm.

This energy falls onto an area of $m^2 * (1 \text{ cm}^2) = 14\,189.8$ cm^2 on the screen. The illuminance E is therefore 0.0596 lm/cm^2, or 55.4 lm/ft^2.

The power radiated by thermal sources is a function of their temperature. We divide thermal sources into two classes, black-body radiators and line sources.

Kirchoff's laws of spectroscopy characterize thermal emission of light from matter.

- A continuous spectrum is emitted by a luminous solid, liquid, or a very dense gas.
- Examples: an incandescent light bulb, glowing coals in the fireplace, element of an

electric heater.
- An emission-line spectrum is emitted by a thin, luminous gas.
- An absorption-line spectrum is produced when white light passes through a cold gas.

Quantum mechanics governs the internal structure of atoms and allows us to elaborate on Kirchoff's Laws.

- Isolated atoms of any element absorb or emit light only at specific wavelengths. Atoms are isolated in a thin gas.
- Every element has its own distinctive pattern of wavelengths at which absorption or emission can occur. The absorption or emission lines which occur in a spectrum therefore allow us to deduce which chemical elements are present.
- The relative strengths of emission or absorption lines of a given element depend on the temperature of the gas absorbing or emitting the light. We therefore have a means of determining the temperature of a gas. By comparing the strengths of emission or absorption lines of different elements, we can deduce quantitatively the relative abundance of each element.
- If we compress a gas, then the internal properties of individual atoms are affected by the electronic properties of neighboring atoms and the absorption or emission lines are no longer sharp, but broadened. The widths of the absorption or emission lines can give us information the density of the gas. If the light emitting material is compressed even more, then all the emission lines blur together completely into a featureless spectrum. This is called a black-body spectrum.

Radiation laws govern the properties of the continuous spectrum.

5.4 Radiation Laws

The primary law governing radiation is the **Planck Radiation Law**, which gives the intensity of radiation emitted by a **blackbody** as a function of wavelength for a fixed temperature. The Planck law gives a distribution, which peaks at some wavelength. The peak shifts to shorter wavelengths for higher temperatures, and the area under the curve grows rapidly with increasing temperature. The diagram below, Figure 5.7, shows the intensity distribution predicted by the Plank law in $J/(m^2 s)$ for blackbodies at various temperature.

Optical Sensing and Measurement

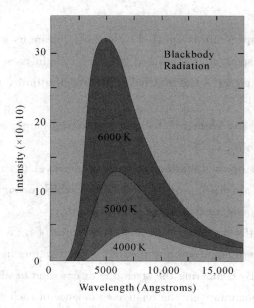

Figure 5.7 Blackbody radiation

$$I(\lambda, T) = \frac{2hc^2}{\lambda^5} \frac{1}{e^{hc/\lambda KT} - 1} \tag{5.7}$$

A blackbody is a body that absorbs all the radiation that falls onto it. It does not reflect any radiation. It reaches thermal equilibrium with its surroundings, and in thermal equilibrium emits exactly as much radiation it absorbs. It has emissivity=1. **Emissivity** measures the fraction of radiative energy that is absorbed by the body.

The **Wien Law** gives the wavelength of the peak of the radiation distribution, $\lambda_{max} = 3 \times 10^6/T$. Here λ is measured in units of nanometer (10^{-9} m) and T is in Kelvin.

The **Stefan-Boltzmann Law** gives the total energy being emitted at all wavelengths by the body.

Radiated power = emissivity $\times \sigma \times T^4 \times$ Area

Here σ is the **Stefan-Boltzmann constant**, $\sigma = 5.67 \times 10^{-8}$ W/(m² · K⁴).

The Wien law explains the shift of the peak to shorter wavelengths as the temperature increases, while the Stefan-Boltzmann law explains the growth in the height of the curve as the temperature increases. This growth is very abrupt, since it varies as the fourth power of the temperature.

Light colored or reflective objects have low emissivity. They do absorb a smaller percentage of the incoming radiation than do dark objects, and also emit radiation less readily.

The Planck radiation law tells us the intensity of the radiation emitted by a hot object as a function of wavelength. The Wien Law gives the wavelength of the peak of the distribution. The surface temperature of the sun is 5800°C = 6073K. The wavelength of the peak of the distribution therefore is 494 nanometer. This wavelength lies in the yellow region of the visible spectrum.

In an incandescent light bulb a filament is heated to approximately 2500 °C = 2773K. This is the maximum temperature that a tungsten filament can stand without evaporating quickly. Compared to the sun, such a filament emits a greater fraction of its radiation in the infrared region of the electromagnetic spectrum. The wavelength of the peak of the distribution is 1082 nanometer. This wavelength lies in the infrared region of the spectrum.

Sunlight and light from an incandescent bulb contain all the colors of the visible spectrum. But the intensity distribution over the different colors is different. Sunlight appears brilliant white while a light bulb looks yellowish.

5.5 Optical Detectors

The detection of optical radiation is usually accomplished by converting the optical energy into an electrical signal. Optical detectors include photon detectors, in which one photon of light energy releases one electron that is detected in the electronic circuitry, and thermal detectors, in which the optical energy is converted into heat, which then generates an electrical signal. Often the detection of optical energy must be performed in the presence of noise sources, which interfere with the detection process. The detector circuitry usually employs a bias voltage and a load resistor in series with the detector. The incident light changes the characteristics of the detector and causes the current flowing in the circuit to change. The output signal is the change in voltage drop across the load resistor. Many detector circuits are designed for specific applications.

5.5.1 Light detection

When light strikes special types of materials, a voltage may be generated, a change in

Optical Sensing and Measurement

electrical resistance may occur, or electrons may be ejected from the material surface. As long as the light is present, the condition continues. It ceases when the light is turned off. Any of the above conditions may be used to change the flow of current or the voltage in an external circuit, and hence may be used to monitor the presence of the light and to measure its intensity.

Detectors suitable for monitoring optical power or energy are commonly employed along with lasers and other light sources. For an application such as laser communication, a detector is necessary as the receiver. For applications involving interferometry, detectors are used to measure the position and motion of the fringes in the interference pattern. In applications involving laser material processing, a detector monitors the laser output to ensure reproducible conditions. In very many applications of light, one desires a detector to determine the output of the laser or other light source. Thus, good optical detectors for measuring optical power and energy are essential.

All optical detectors respond to the power in the optical beam, which is proportional to the square of the electric field. They are thus called "square-law detectors." Microwave detectors, in contrast, can measure the electric field intensity directly. But all the detectors that we consider here exhibit square-law response. This is also true of other common optical detectors such as the human eye and photographic film.

The detection and measurement of optical and infrared radiation is a well-established area of technology. In recent years, this technology has been applied specifically to laser applications, and detectors particularly suitable for use with lasers have been developed. Commercial developments have also kept pace. Detectors specially designed and packaged for use with lasers are marketed by numerous manufacturers. Some detectors are packaged in the format of a power or energy meter. These devices include a complete system for measuring the output of a specific class of lasers, and include a detector, housing, amplification if necessary, and a readout device.

In this module, we will describe some of the detectors that are available. We shall not attempt to cover the entire field of light detection, which is very broad. Instead, we shall emphasize those detectors that are most commonly encountered. We shall also define some of the common terminology.

There are two broad classes of optical detectors: photon detectors and thermal detectors. Photon detectors rely on the action of quanta of light energy to interact with electrons in the detector material and to generate free electrons. To produce such effects, the quantum of light must have sufficient energy to free an electron. The wavelength response of photon detectors shows a long-wavelength cutoff. When the wavelength is longer than the cutoff wavelength, the photon energy is too small to liberate an electron and the response of the detector drops to zero.

Thermal detectors respond to the heat energy delivered by the light. The response of these detectors involves some temperature-dependent effect, like a change of electrical resistance. Because thermal detectors rely on only the amount of heat energy delivered, their response is independent of wavelength.

The output of photon detectors and thermal detectors as a function of wavelength is shown schematically in Figure 5.8. This shows how the output of thermal detectors is independent of wavelength. It also shows the typical spectral dependence of the response of photon detectors, which increases with increasing wavelength until the cutoff wavelength is reached. At that point it drops rapidly to zero.

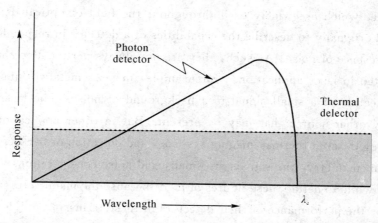

Figure 5.8 Schematic drawing of the output of photon detectors and thermal detectors as a function of wavelength. The position of the long-wavelength cutoff, λ_c, for photon detectors is indicated.

Photon detectors may be further subdivided into the following groups:
- Photoconductive. The electrical conductivity of the material changes as a function of the intensity of the incident light. Photoconductive detectors are semiconductor materials. They have an external electrical bias voltage.
- Photovoltaic. These detectors contain a p-n semiconductor junction and are often called photodiodes. A voltage is self generated as radiant energy strikes the device. The

photovoltaic detector may operate without external bias voltage. A good example is the solar cell used on spacecraft and satellites to convert the sun's light into useful electrical power.

- Photoemissive. These detectors use the photoelectric effect, in which incident photons free electrons from the surface of the detector material. These devices include vacuum photodiodes, bipolar phototubes, and photomultiplier tubes.

Photoconductive and photovoltaic detectors are commonly used in circuits in which there is a load resistance in series with the detector. The output is read as a change in the voltage drop across the resistor.

We shall describe these effects in more detail later in the discussion of the types of detectors.

5.5.2 Detector characteristics

The performance of optical detectors is commonly described by a number of different figures of merit, which are widely used throughout the field of optical detection. They were developed originally to describe the capabilities of a detector in responding to a small signal in the presence of noise. As such, they are not always pertinent to the detection of laser light. Often in laser applications—for example, in laser metalworking—there is no question of detection of a small signal in a background of noise. The laser signal is far larger than any other source that may be present. But in other applications, like laser communications, infrared thermal imaging systems, and detection of backscattered light in laser Doppler anemometry, the signals are small, and noise considerations are important. It is also worthwhile to define these figures of merit because the manufacturers of detectors usually describe the performance of their detectors in these terms.

The first term that is commonly used is responsivity. This is defined as the detector output per unit of input power. The units of responsivity are either amperes/watt (alternatively milliamperes/milliwatt or microamperes/microwatt, which are numerically the same) or volts/watt, depending on whether the output is an electric current or a voltage. This depends on the particular type of detector and how it is used. Figure 5.8 can be considered to be a representation of how the responsivity varies with wavelength for photon detectors and thermal detectors. We note that the responsivity gives no information about noise characteristics.

Chapter 5 Sources and Detectors

The responsivity is an important characteristic that is usually specified by the manufacturer, at least as a nominal value. Knowledge of the responsivity allows the user to determine how much detector signal will be available in a specific application. One may also characterize the spectral responsivity, which is the responsivity as a function of wavelength.

A second figure of merit, one that depends on noise characteristics, is the noise equivalent power (NEP). This is defined as the radiant power that produces a signal voltage (current) equal to the noise voltage (current) of the detector. Since the noise is dependent on the bandwidth of the measurement, that bandwidth must be specified. The equation defining NEP is

$$\text{NEP} = \frac{IAV_N}{V_S(\Delta f)^{\frac{1}{2}}} \qquad (5.8)$$

where I is the irradiance incident on the detector of area A, V_N is the root mean square noise voltage within the measurement bandwidth Δf, and V_S is the root mean square signal voltage. The NEP has units of watts per (hertz to the one-half power), commonly called watts per root hertz. From the definition, it is apparent that the lower the value of the NEP, the better are the characteristics of the detector for detecting a small signal in the presence of noise.

The NEP of a detector is dependent on the area of the detector. To provide a figure of merit under standard conditions, a term called detectivity is defined. Detectivity is represented by the symbol $D*$, which is pronounced as D-star. It is defined by

$$D* = \frac{A^{1/2}}{\text{NEP}} \qquad (5.9)$$

Since many detectors have NEP proportional to the square root of their area, $D*$ is independent of the area of the detector and provides a measure of the intrinsic quality of the detector material itself, independent of the area with which the detector happens to be made. When a value of $D*$ for a photodetector is measured, it is usually measured in a system in which the incident light is modulated or chopped at a frequency f so as to produce an AC signal, which is then amplified with an amplification bandwidth Δf. The dependence of $D*$ on the wavelength λ, the frequency f at which the measurement is made, and the bandwidth f are specified in the notation $D*(\lambda, f, \Delta f)$. The reference bandwidth is frequently taken as 1 hertz. The units of $D*(\lambda, f, \Delta f)$ are centimeters (square root hertz) per watt. A high value of $D*(\lambda, f, \Delta f)$ means that the detector is suitable for detecting weak signals in the presence of noise. Later, in the discussion of noise, we will describe the effect of the modulation frequency and the bandwidth on the noise characteristics.

Another commonly encountered figure of merit for photodetectors is the quantum efficiency. Quantum efficiency is defined as the ratio of countable events produced by photons incident on the detector to the number of photons. If the detector is a photoemissive detector that emits free electrons from its surface when light strikes it, the quantum efficiency is the number of free electrons divided by the number of incident photons. If the detector is a semiconductor p-n junction device in which hole-electron pairs are produced, the quantum efficiency is the number of hole-electron pairs divided by the number of incident photons. Thus if, over a period of time, 100,000 photons are incident on the detector and 10,000 hole-electron pairs are produced, the quantum efficiency is 10%.

The quantum efficiency is basically another way of expressing the effectiveness of the incident radiant energy for producing electrical current in a circuit. It may be related to the responsivity by the equation:

$$Q = 100 R_d h v = 100 R_d \left(\frac{1.2395}{\lambda} \right) \qquad (5.10)$$

where Q is the quantum efficiency (in %) and R_d is the responsivity (in amperes per watt) of the detector at wavelength λ (in micrometers), and hv is the photon energy. The right portion of the equation makes calculation easy for a specific responsivity at a given wavelength.

Another important detector characteristic is the speed of the detector response to changes in light intensity. If a constant source of light energy is instantaneously turned on and irradiates a photodetector, it will take a finite time for current to appear at the output of the device and for the current to reach a steady value. If the same source is turned off instantaneously, it will take a finite time for the current to decay back to its initial zero value. The term response time generally refers to the time it takes the detector current to rise to a value equal to 63.2% of the steady-state value reached after a relatively long period of time. (This value is numerically equal to $\bar{l} - 1/e$, where e is the base of the natural logarithms.) The recovery time is the time photocurrent takes to fall to 36.8% of the steady-state value when the light is turned off instantaneously.

Because photodetectors often are used for detection of fast pulses, a more important term, called rise time, is often used to describe the speed of the detector response. Rise time is defined as the time difference between the points at which the detector has reached 10% of its peak output and the point at which it has reached 90% of its peak response, when it is irradiated by a very short pulse of light. The fall time is defined as the time between the 90% point and the 10% point on the trailing edge of the pulse waveform. This is also called the decay time. We note that the fall time may be different numerically from the rise time.

Of course light sources are not turned on or off instantaneously. For accurate measurements of rise time and fall time, the source used for the measurement should have a rise time much less that the rise time of the detector that is being tested. Generally one will accept a source whose rise time is less than 10% of the rise time of the detector being tested.

Other factors that affect measured rise times are the limitations introduced by the electrical cables and by the display device, for example, the oscilloscope or recorder. These devices can make the measured rise time appear longer than the value that arises from the detector alone.

The response time of a photodetector arises from the transit time of photogenerated charge carriers within the detector material and from the inherent capacitance and resistance associated with the device. It is also affected by the value of the load resistance that is used with the detector. There is a tradeoff in the selection of a load resistance between speed of response and high sensitivity. It is not possible to achieve both simultaneously. Fast response requires a low load resistance (generally 50 ohms or less), whereas high sensitivity requires a high value of load resistance. It is also important to keep any capacitance associated with the circuitry or display device as low as possible. This will help to keep the RC time constant low.

Manufacturers often quote nominal values for the rise times of their detectors. These should be interpreted as minimum values, which may be achieved only with careful circuit design and avoidance of excess capacitance and resistance.

Another important characteristic of detectors is their linearity. Photodetectors are characterized by a response that is linear with incident intensity over a broad range, perhaps many orders of magnitude. If the output of the detector is plotted versus the input power, there should be no change in the slope of the curve. Noise will determine the lowest level of incident light that is detectable. The upper limit of the input/output linearity is determined by the maximum current that the detector can handle without becoming saturated. Saturation is a condition in which there is no further increase in detector response as the input light is increased. Linearity may be quantified in terms of the maximum percentage deviation from a straight line over a range of input light levels. For example, the maximum deviation from a straight line could be 5% over the range of input light from 10^{-12} W/cm² to 10^{-4} W/cm². One would state that the linearity is 5% over eight orders of magnitude in the input.

The manufacturer often specifies a maximum allowable continuous light level. Light levels in excess of this maximum may cause saturation, hysteresis effects, and irreversible damage to the detector. If the light occurs in the form of a very short pulse, it may be possible to exceed the continuous rating by some factor (perhaps as much as 10 times) without damage or noticeable changes in linearity.

5.5.3 Noise considerations

The topic of noise in optical detectors is a complex subject. In this module we will do no more than present some of the most basic ideas. Noise is defined as any undesired signal. It masks the signal that is to be detected. Noise can be distinguished as external and internal. External noise involves those disturbances that appear in the detection system because of actions outside the system. Examples of external noise could be pickup of hum induced by 60-Hz electrical power lines and static caused by electrical storms. Internal noise includes all noise generated within the detection system itself. Every electronic device has internal sources of noise, which may be considered as an ever-present limit to the smallest signal that may be handled by the system.

Noise cannot be described in the same manner as usual electric currents or voltages. We think of currents or voltages as functions of time, such as constant direct currents or sine-wave alternating voltages. The noise output of an electrical circuit as a function of time is completely erratic. We cannot predict what the output will be at any instant. There will be no indication of regularity in the waveform. The output is said to be random.

The output from a random noise generator might look like what is shown in Figure 5.9, which is a plot of instantaneous voltage as a function of time. Because of the random nature of the noise, the voltage fluctuates about an average value V_{av}. How does one describe these variations? A simple average is meaningless because the average is zero.

Rather, one uses an average of the squares of the deviations around V_{av}, with the average taken over a period of time T much longer than the period of the fluctuations.

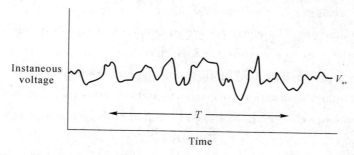

Figure 5.9 A record of random noise voltage

Mathematically this is expressed as:

$$\overline{V^2} = \overline{(V(t) - V_{av})^2} = \frac{1}{T}\int_0^T (V(t) - V_{av})^2 \, dt \tag{5.11}$$

where $V(t)$ is the value of the voltage at time t and $\overline{V^2}$ is termed the mean square voltage fluctuation, and the bar over a quantity indicates an average value.

The right side of the equation contains an integral sign with limits 0 and T. This means that one adds all values $(V(t) - V_{av})^2$ for each small increment of time dt in the interval from time 0 to time T. It is unlikely that one would calculate the noise voltage directly from this equation, because the process would be laborious. The equation does define the basic concept.

If two or more noise sources are present, their total effect is found by adding their mean square voltages. Since mean square voltages are proportional to power, this is equivalent to saying that noise powers are additive, but noise voltages or currents are not. As an example, if two independent noise sources are present in a circuit, with root-mean-square (rms) noise voltages of 30 and 40 microvolts, the total rms noise voltage is $(30^2 + 40^2)^{1/2}$ or 50 microvolts.

Now we will consider some of the sources of noise encountered in optical detector applications. A complete description of all types of noise would be very long. We will describe the four types most often encountered in a system for visible and infrared detection:

- Johnson noise
- Shot noise
- $1/f$ noise

Optical Sensing and Measurement

- Photon noise

Johnson noise is a type of noise generated by thermal fluctuations in conducting materials. It is sometimes called thermal noise. It results from the random motion of electrons in a conductor. The electrons are in constant motion, colliding with each other and with the atoms of the material. Each motion of an electron between collisions represents a tiny current. The sum of all these currents taken over a long period of time is zero, but their random fluctuations over short intervals constitute Johnson noise.

The mean square value of the voltage associated with Johnson noise is:

$$\overline{V^2} = 4KTR\Delta f \qquad (5.12)$$

where K is Boltzmann's constant (1.38×10^{-23} joule/degree Kelvin), T is the absolute temperature, R is the circuit resistance (ohms), and Δf is the bandwidth of the amplification (Hz). Since the load resistance is usually greater than the internal resistance of the photodetector, the Johnson noise may be dominated by the load resistor. This equation is also frequently written as $V^2 = 4KTRB$, with B denoting the bandwidth.

The equation indicates methods to reduce the magnitude of the Johnson noise. One may cool the system, especially the load resistor. One should reduce the value of the load resistance, although this is done at the price of reducing the available signal. One should keep the bandwidth of the amplification small; one Hz is a commonly employed value.

The term shot noise is derived from fluctuations in the stream of electrons in a vacuum tube. These variations create noise because of the random fluctuations in the arrival of electrons at the anode. It originally was likened to the the noise of a hail of shot striking a target; hence the name shot noise was applied. In semiconductors, the major source of noise is due to random variations in the rate at which charge carriers are generated and recombine. This noise, called generation-recombination or gr noise, is the semiconductor counterpart of shot noise. The mean-square current fluctuation for shot noise in a semiconductor photodetector is:

$$\overline{i^2} = \frac{2eI_{DC}}{f} \qquad (5.13)$$

where e is the electronic charge (1.6×10^{-19} coulomb), I_{DC} is the DC component of dark leakage current (or any current) in amperes, and Δf is the bandwidth of the amplification (Hz). Note that here we have specified a mean square noise current, instead of a voltage.

The shot noise may be minimized by keeping any DC component to the current small, especially the dark current, and by keeping the bandwidth of the amplification system small.

The term $1/f$ noise (pronounced one over f) is used to describe a number of types of noise that are present when the modulation frequency f is low. This type of noise is also called excess noise because it exceeds shot noise at frequencies below a few hundred Hertz. With respect to photodiodes, it is sometimes called boxcar noise, because it can suddenly appear and then disappear in small boxes of noise observed over a period of time.

The mechanisms that produce $1/f$ noise are poorly understood and there is no simple mathematical expression to define $1/f$ noise. The noise power is inversely proportional to f, the modulation frequency. This dependence of the noise power leads to the name for this type of noise.

To reduce $1/f$ noise, a photodetector should be operated at a reasonably high frequency, often taken as 1000 Hz. This is a high enough value to reduce the contribution of $1/f$ noise to a small amount. Since to reduce Johnson noise and shot noise, the amplification bandwidth should be small (perhaps 1 Hz), measurements of the spectral detectivity are often expressed as $D*(\lambda, 1000, 1)$.

Even if all the previously discussed sources of noise could be eliminated, there would still be some noise present in the output of a photodetector because of the random arrival rate of photons from the source of radiant energy that is being measured and from the background. This contribution to the noise is called photon noise, and is a noise source external to the detector. It imposes the ultimate fundamental limit to the detectivity of a photodetector.

The photon noise associated with the fluctuations in the arrival rate of photons in the desired signal is not something that can be reduced. The contribution of fluctuations in the arrival of photons from the background, a contribution that is called background noise, can be reduced. The background noise increases with the field of view of the detector and with the temperature of the background. In some cases it may be possible to reduce the field of view of the detector so as to view only the source of interest. In other cases it may be possible to cool the background. Both these measures may be used to reduce the background noise contribution to photon noise.

Any of the types of noise described here, or a combination of the noise powers from a combination of the sources, will set an upper limit to the detectivity of a photodetector system.

5.5.4 Types of detectors

Photon Detectors

We have previously distinguished photon detectors and thermal detectors. We begin the detailed discussion of detector types with photon detectors. As mentioned before, photon detectors rely on liberating free electrons and require the photon to have sufficient energy to exceed some threshold, i. e., the wavelength must be shorter than the cutoff wavelength. We will consider three types of photoeffect that are often used for detectors. These are the photovoltaic effect, the photoemissive effect, and the photoconductive effect.

The photovoltaic effect and the operation of photodiodes both rely on the presence of a p-n junction in a semiconductor. When such a junction is in the dark, an electric field is present internally in the junction region because there is a change in the level of the conduction and valence bands. This change leads to the familiar electrical rectification effect produced by such junctions.

When light falls on the junction, it is absorbed and, if the photon energy is large enough, it produces free hole-electron pairs. The electric field at the junction separates the pair and moves the electron into the n-type region and the hole into the p-type region. This leads to a change in voltage that may be measured externally. This process is the origin of the so-called photovoltaic effect. The photovoltaic effect is the generation of a voltage when light strikes a semiconductor p-n junction. We note that the open-circuit voltage generated in the photovoltaic effect may be detected directly and that no bias voltage or ballast resistor is required.

It is also possible to use a p-n junction to detect light if one does apply a bias voltage in the reverse direction. By reverse direction, we mean the direction of low current flow, that is, with the positive voltage applied to the n-type material. A p-n junction detector with bias voltage is termed a photodiode. The current-voltage characteristics of a photodiode are shown in Figure 5.10. The curve labeled "dark" represents conditions in the absence of light; it displays the familiar rectification characteristics of a p-n semiconductor diode. The other curves show the current-voltage characteristics when the device is illuminated at different light levels. The characteristics of a photovoltaic detector, with zero applied voltage, are represented by the intersections of the different curves with the vertical axis.

Chapter 5 Sources and Detectors

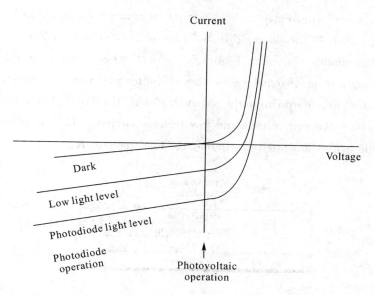

Figure 5.10 Current voltage characteristics for photovoltaic detectors and photodiodes. Regions for photodiode and photovoltaic operation are indicated.

The photodiode detector is operated in the lower left quadrant of this figure. The current that may be drawn through an external load resistor increases with increasing light level. In practice, one measures the voltage drop appearing across the resistor.

To increase the frequency response of photodiodes, a type called the PIN photodiode has been developed. This device has a layer of nearly intrinsic material bounded on one side by a relatively thin layer of highly doped p-type semiconductor, and on the other side by a relatively thick layer of n-type semiconductor. A sufficiently large reverse bias voltage is applied so that the depletion layer, from which free carriers are swept out, spreads to occupy the entire volume of intrinsic material. This volume then has a high and nearly constant electric field. It is called the depletion region because all mobile charges have been removed. Light that is absorbed in the intrinsic region produces free electron-hole pairs, provided that the photon energy is high enough. These carriers are swept across the region with high velocity and are collected in the heavily doped regions. The frequency response of such PIN photodiodes can be very high, of the order of 10^{10} Hz. This is higher than the frequency response of p-n junctions without the intrinsic region.

A variety of photodiode structures is available. No single photodiode structure can best meet all system requirements. Therefore, a number of different types has been developed. These include the planar diffused photodiode, shown in Figure 5.11(a), and the Schottky photodiode, shown in Figure 5.11(b). The planar diffused photodiode is formed by growing a layer of oxide over a slice of high-resistivity silicon, etching a hole in the oxide, and diffusing boron into the silicon through the hole. This structure leads to devices with high breakdown voltage and low leakage current. The circuitry for operation of the photodiode is also indicated, including the load resistor R_L.

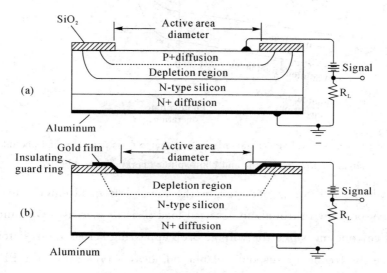

Figure 5.11 Photodiode structures
(a) Planar diffused photodiode; (b) Schottky photodiode. The load resistor is denoted R_L.

The Schottky barrier photodiode uses a junction between a metallic layer and a semiconductor. If the metal and the semiconductor have work functions related in the proper way, this can be a rectifying barrier. The junction is formed by oxidation of the silicon surface, etching of a hole in the oxide, and then evaporation of a thin transparent and conducting gold layer. The insulation guard rings serve to reduce the leakage current through the devices.

A number of different semiconductor materials is in common use as optical detectors. They include silicon in the visible and near ultraviolet and near infrared, germanium and indium gallium arsenide in the near infrared, and indium antimonide, indium arsenide, mercury cadmium telluride, and germanium doped with elements like copper and gold in the longer-wavelength infrared.

The most frequently encountered type of photodiode is silicon. Silicon photodiodes are widely used as the detector elements in optical disks and as the receiver elements in optical-fiber telecommunication systems operating at wavelengths around 800 nm. Silicon photodiodes respond over the approximate spectral range of 400-1100 nm, covering the visible and part of the near-infrared regions. The spectral responsivity of typical commercial silicon photodiodes is shown in Figure 5.12. The responsivity reaches a peak value around 0.7 amp/watt near 900 nm, decreasing at longer and shorter wavelengths. Optional models provide somewhat extended coverage in the infrared or ultraviolet regions. Silicon photodiodes are useful for detection of many of the most common laser wavelengths, including argon, He-Ne, AlGaAs, and Nd:YAG. As a practical matter, silicon photodiodes have become the detector of choice for many laser applications. They represent well-developed technology and are widely available. They represent the most widely used type of laser detectors for lasers operating in the visible and near-infrared portions of the spectrum.

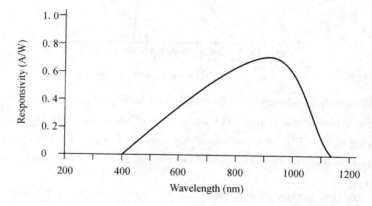

Figure 5.12 Responsivity as a function of wavelength for typical silicon photodiodes

Other types of photodetectors are available for detection of light in other wavelength regions. Photodetectors have been fabricated from many other materials. Different detector materials are useful in different spectral regions. Figure 5.13 shows the spectral $D*$ (or detectivity) for a number of commercially available detectors operating in the infrared spectrum. The figure includes both photovoltaic detectors and photoconductive detectors, which will be described in more detail later. The choice of detector will depend on the wavelength region that is desired. For example, for a laser operating at 5 mm, an indium antimonide photovoltaic detector would be suitable.

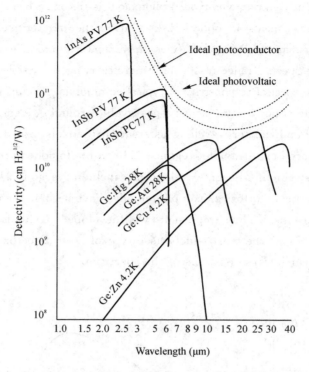

Figure 5.13 Detectivity as a function of wavelength for a number of different types of photodetectors operating in the infrared spectrum. The temperature of operation is indicated. Photovoltaic detectors are denoted PV; photoconductive detectors are denoted PC. The curves for ideal photodetectors assume a 2p steradian field of view and a 295-K background temperature.

The shape of the curves in Figure 5.13 is characteristic of photon detectors. One photon produces one electron-hole pair in the material, so long as the photon energy is high enough. Absorption of one photon then gives a constant response, independent of wavelength (provided that the wavelength lies within the range of spectral sensitivity of the detector). One photon of ultraviolet light and one photon of infrared light each produces the same result, even though they have much different energy. For constant photon arrival rate, as wavelength increases, the incident power decreases, but the response remains the same. Therefore, the value of $D*$ increases, reaching a maximum at the cutoff wavelength, which is equal to the Planck's constant times the velocity of light divided by the band gap of the material. At longer wavelengths, the detectivity decreases rapidly because the photons do not have enough energy to excite an electron into the conduction band.

Figure 5.13 also indicates the detectivity for "ideal" detectors, that is, detectors whose performance is limited only by fluctuations in the background of incident radiation, and that do not contribute noise themselves. Available detectors approach the ideal performance limits fairly closely.

Another variation of the photodiode is the avalanche photodiode. The avalanche photodiode offers the possibility of internal gain; it is sometimes referred to as a "solid-state photomultiplier." The most widely used material for avalanche photodiodes is silicon, but they have been fabricated from other materials, such as germanium. An avalanche photodiode has a diffused p-n junction, with surface contouring to permit high reverse-bias voltage without surface breakdown. A large internal electric field leads to multiplication of the number of charge carriers through ionizing collisions. The signal is thus increased, to a value perhaps 100-200 times greater than that of a nonavalanche device. The detectivity is also increased, provided that the limiting noise is not from background radiation. Avalanche photodiodes cost more than conventional photodiodes, and they require temperature-compensation circuits to maintain the optimum bias, but they represent an attractive choice when high performance is required.

Silicon and other photodiodes have been configured as power meters, which are calibrated so that the detector output may be presented on a display as the laser power. These devices may be used directly to measure the power of a continuous laser or of a repetitively pulsed laser operating at a reasonably high pulse-repetition rate. Some commercial units can measure powers down to the nanowatt regime. Because the photodiode response varies with wavelength, the manufacturer usually supplies a calibration graph to allow conversion to wavelengths other than that for which the photodiode was calibrated. Alternatively, some units are supplied with filters to compensate for the wavelength variation of the detector response, so that the response of the entire unit will be wavelength-independent, at least over some interval. Such packaged power meters provide a convenient and useful monitor of laser output power.

A photoemissive detector employs a cathode coated with a material that emits electrons when light of wavelength shorter than a certain value falls on the surface. The electrons emitted from the surface may be accelerated by a voltage to an anode where they give rise to a current in an external circuit. These detectors are available commercially from a number of manufacturers. They represent an important class of detectors for many applications.

Some standardized spectral response curves for photoemissive cathodes are shown in Figure 5.14. The materials have low work functions, that is, incident light may easily cause the surfaces to emit an electron. The cathodes are often mixtures containing alkali metals, such as sodium and potassium. The usefulness of these devices ranges from the ultraviolet to the near infrared. At wavelengths longer than 1.2 μm, no photoemissive response is available. The short-wavelength end of the response curve is set by the nature of the window material used in the tube that contains the detector. The user can select a device with a cathode with maximum response in a selected wavelength region.

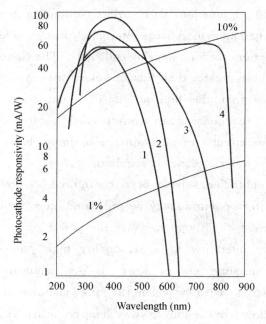

Figure 5.14 Response as a function of wavelength for a number of photoemissive surfaces. Curve 1 is the response of a bialkali-type cathode with a sapphire window; curve 2 is for a different bialkali cathode with a lime glass window; curve 3 is for a multialkali cathode with a lime glass window; and curve 4 is for a GaAs cathode with a 9741 glass window. The curves beled 1 and 10% denote what the response would be at the indicated value of quantum efficiency.

An important variation of the photoemissive detector is the photomultiplier. This is a device with a photoemissive cathode and a number of secondary emitting stages called dynodes. The dynodes are arranged so that electrons from each dynode are delivered to the next dynode in the series. Electrons emitted from the cathode are accelerated by an applied voltage to the first dynode, where their impact causes emission of numerous secondary electrons. These electrons are accelerated to the next dynode and generate even more electrons. Finally, electrons from the last dynode are accelerated to the anode and produce a large current pulse in the external circuit. The photomultiplier is packaged as a vacuum tube.

Figure 5.15 shows a cross-sectional diagram of a typical photomultiplier tube structure. This tube has a transparent end window with the underside coated with the photocathode material. Figure 5.16 shows the principles of operation of the tube. With careful design, photoelectrons emitted from the cathode will strike the first dynode, where they produce 1 to 8 secondary electrons per incident electron. These are accelerated to the second dynode, where the process is repeated. After several such steps the electrons are collected at the anode and flow through the load resistor. Voltages of 100 to 300 volts are required to accelerate electrons between dynodes, so that the total tube voltage may be from 500 to 3000 volts from anode to cathode, depending on the number of dynodes.

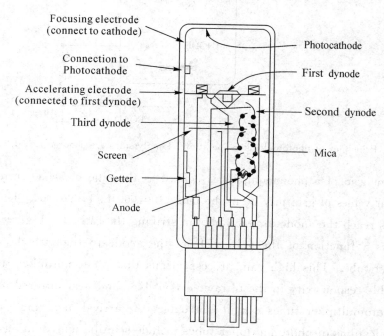

Figure 5.15 Diagram of typical photomultiplier tube structure

Optical Sensing and Measurement

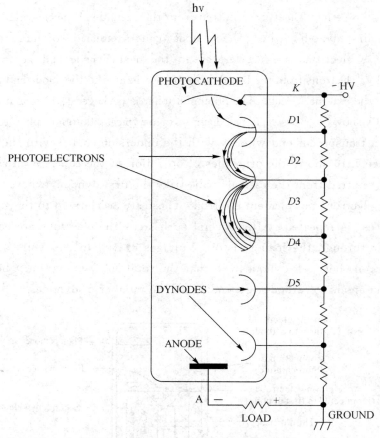

Figure 5.16 Principles of photomultiplier operation. The dynodes are denoted $D1$, $D2$, and so on.

The current gain of a photomultiplier is defined as the ratio of anode current to cathode current. Typical values of gain may be in the range 100,000 to 1,000,000. Thus 100,000 or more electrons reach the anode for each photon striking the cathode. Figure 5.17 shows a plot of gain as a function of the voltage from the anode to the cathode, for a typical photomultiplier tube. This high gain process means that photomultiplier tubes offer the highest available responsivity in the ultraviolet, visible, and near-infrared portions of the spectrum. Photomultiplier tubes can in fact detect the arrival of a single photon at the cathode. Applications of photomultiplier tubes include scintillation counting, air-pollution monitoring, photon counting, star tracking, photometry, and radiometry.

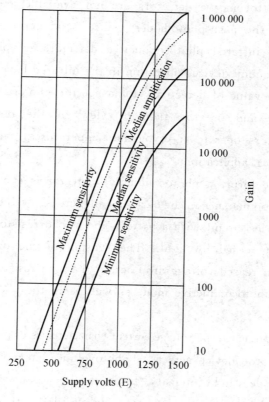

Figure 5.17 Photomultiplier gain as a function of applied voltage

A third class of photodetector uses the phenomenon of photoconductivity. A semiconductor in thermal equilibrium contains free electrons and holes. The concentration of electrons and holes is changed if light is absorbed by the semiconductor. The light must have photon energy large enough to cause excitation, either by raising electrons across the forbidden band gap or by activating impurities present within the band gap. The increased number of charge carriers leads to an increase in the electrical conductivity of the semiconductor. The device is used in a circuit with a bias voltage and a load resistor in series with it. The change in electrical conductivity leads to an increase in the current flowing in the circuit, and hence to a measurable change in the voltage drop across the load resistor.

Photoconductive detectors are most widely used in the infrared spectrum, at

Optical Sensing and Measurement

wavelengths where photoemissive detectors are not available and the wavelengths are beyond the cutoffs of the best photodiodes (silicon and germanium). Many different materials are used as infrared photoconductive detectors. Typical values of spectral detectivity for some common devices operating in the infrared have already been shown in Figure 5.13 The exact value of detectivity for a specific photoconductor depends on the operating temperature and on the field of view of the detector. Most infrared photoconductive detectors operate at cryogenic temperatures, which may involve some inconvenience in practical applications.

In its most simple form, a photoconductive detector is a crystal of semiconductor material that has low conductance in the dark and an increased value of conductance when it is illuminated. In a series circuit with a battery and a load resistor, the detector element has its conductance increased by light. The sensing of the presence of the light is accomplished via the increased voltage drop across the load resistor. But it is possible to use photodiodes in a photoconductive mode as well as in the photovoltaic mode that we have already described.

In Figure 5.10 we presented the voltage-current characteristics of p-n junction photodetectors. In the first quadrant, the device acts as a photovoltaic detector. It produces a voltage proportional to the incident light intensity. In the third quadrant, when a reverse voltage is applied, it acts as a photoconductive detector. In the dark, the reverse current is very small. When light strikes the diode, there is little increase in the forward current, but the reverse current can increase significantly. The current is proportional to the intensity of the incident light, and the increase can be linear over many orders of magnitude.

Figure 5.18 shows the voltage-current characteristics in a different way. Note that the curve has been rotated 180 degree from Figure 5.10. The photoconductive mode of operation appears on the right side of the figure. Negative values of reverse voltage increase from left to right, starting at the origin. The forward-biased (photovoltaic) device is a voltage generator. If it operates with a low-resistance load, the operation is along the near-vertical line and the current output is fairly linear with input radiation. As the load resistance increases, the output becomes nonlinear. If one observes the open-circuit voltage (load line horizontal), the open-circuit voltage is found to be proportional to the logarithm of the input light intensity.

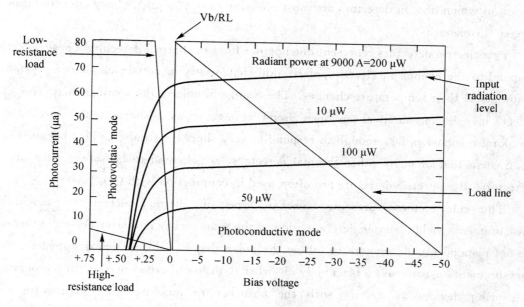

Figure 5.18 Volt-ampere characteristics of photodiodes

In the reverse-biased or photoconductive mode, linear operation is maintained so long as the photodiode is not saturated and the bias voltage is higher than the product of the load resistance and the current.

For a reverse-biased device, the photodiode exhibits higher responsivity, faster response time, and greater linearity than when operated in the forward-biased mode. One drawback is the presence of a small dark current. In the forward-biased mode, the dark current may be eliminated. This makes photovoltaic devices desirable for low-level measurements in which the dark current would interfere. But the responsivity and speed decrease and the response becomes nonlinear for large values of load resistance.

Thermal detectors

Now we turn to the second broad class of photodetectors, thermal detectors. Thermal detectors respond to the total energy absorbed, regardless of wavelength. They have no long-wavelength cutoff in their response, as photon detectors do. The value for $D*$ for a thermal detector is independent of wavelength. Thermal detectors generally do not offer as rapid response as photon detectors, and for laser work are not often used in the wavelength

Optical Sensing and Measurement

region in which photon detectors are most effective ($\lambda = 1.55$ μm). They are often used at longer wavelengths.

Pyroelectric detectors represent one popular form of thermal detector. These detectors respond to the change in electric polarization that occurs in certain classes of crystalline materials as their temperature changes. The change in polarization, called the pyroelectric effect, may be measured as an open-circuit voltage or as a short-circuit current. The temporal response is fast enough to respond to very short laser pulses. This behavior is in contrast to that of many other thermal detectors, which tend to be slower than photon detectors. Pyroelectric detectors are often used in conjunction with CO_2 lasers.

The calorimeter represents another type of thermal detector. Calorimetric measurements yield a simple determination of the total energy in a laser pulse, but usually do not respond rapidly enough to follow the pulse shape. Calorimeters designed for laser measurements usually use a blackbody absorber with low thermal mass with temperature-measuring devices in contact with the absorber to measure the temperature rise. Knowledge of the thermal mass coupled with measurement of the temperature rise yields the energy in the laser pulse. The temperature-measuring devices include thermocouples, bolometers, and thermistors. Bolometers and thermistors respond to the change in electrical resistivity that occurs as temperature rises. Bolometers use metallic elements; thermistors use semiconductor elements.

Many different types of calorimeters have been developed for measuring the total energy in a laser pulse or for integrating the output from a continuous laser. Since the total energy in a laser pulse is usually not large, the calorimetric techniques are rather delicate. The absorbing medium must be small enough that the absorbed energy may be rapidly distributed throughout the body. It must be thermally isolated from its surroundings so that the energy is not lost.

One form of calorimeter uses a small, hollow carbon cone, shaped so that radiation entering the base of the cone will not be reflected back out of the cone. Such a design acts as a very efficient absorber. Thermistor beads or thermocouples are placed intimately in contact with the cone. The thermistors form one element of a balanced bridge circuit, the output of which is connected to a display or meter. As the cone is heated by a pulse of energy, the resistance of the bridge changes, leading to an unbalance of the bridge and a

Chapter 5 Sources and Detectors

voltage pulse that activates the display. The pulse decays as the cone cools to ambient temperature. The magnitude of the voltage pulse gives a measure of the energy in the pulse. In some designs, two identical cones are used to form a conjugate pair in the bridge circuit. This approach allows cancellation of drifts in the ambient temperature.

A calorimeter using a carbon cone or similar design is a simple and useful device for measurement of laser pulse energy. In the range of energy below 1 J or so, an accuracy of a few percent or better should be attainable. The main sources of error in conical calorimeters for pulsed energy measurements are loss of some of the energy by reflection, loss of heat by cooling of the entire system before the heat is distributed uniformly, and imperfect calibration. Calibration using an electrical-current pulse applied to the calorimeter element can make the last source of error small. With careful technique, the other sources of error can be held to a few percent.

Calorimeters using absorbing cones or disks with thermocouples to sense the temperature rise have been developed for laser pulses with energy up to hundreds of joules. When the laser energy becomes high, destructive effects, such as vaporization of the absorbing surface, may limit the usefulness of calorimeters. Since calorimeters require surface absorption of the laser energy, there are limits to the energy that a calorimeter can withstand without damage.

If the response of the calorimeter is fast, it can be used for measurement of power in a continuous laser beam. The temperature of the absorber will reach an equilibrium value dependent on the input power. Such units are available commercially as laser power meters, with different models capable of covering the range from fractions of a milliwatt to ten kilowatts.

Compared to the power meters based on silicon or other photodiodes, the power meters based on absorbing cones or disks are useful over a wider range of wavelength, and do not require use of a compensating factor to adjust for the change in response as the laser wavelength changes. Also, power meters based on these thermal detectors tend to cover a higher range of laser power than do the models based on photodiodes.

Some values of $D*$ for thermal detectors are shown in Figure 5.19. The values are independent of wavelength. In the visible and near infrared, the values of $D*$ for thermal detectors tend to be lower than for good photon detectors, but the response does not decrease at long wavelength.

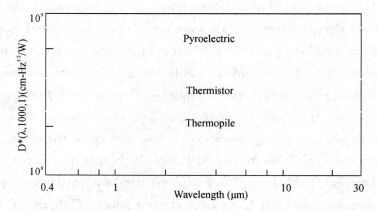

Figure 5.19 Detectivity ($D*$) as a function of wavelength for several typical thermal detectors. The temperature of operation is 295 K.

5.5.5 Calibration

The response of any photodetector in current (or voltage) per unit input of power is often taken as the nominal value specified by the manufacturer. For precise work, the detector may have to be calibrated by the user. But accurate absolute measurements of power or energy are difficult. A good calibration requires painstaking work.

Quantitative measurements of laser output involve several troublesome features. The intense laser output tends to overload and saturate the output of detectors if they are exposed to the full power. Thus, absorbing filters may be used to cut down the input to the detector. A suitable filter avoids saturation of the detector, keeps it in the linear region of its operating characteristics, shields it from unwanted background radiation, and protects it from damage. Many types of attenuating filters have been used, including neutral-density filters, semiconductor wafers (like silicon), and liquid filters. Gelatin or glass neutral-density filters and semiconductor wafers are subject to damage by high-power laser beams. Liquid filters containing a suitable absorber (for example, an aqueous solution of copper sulfate for ruby lasers) are not very susceptible to permanent damage.

The calibration of filters is a difficult task, because the filters also saturate and become nonlinear when exposed to high irradiance. If a certain attenuation is measured for a filter exposed to low irradiance, the attenuation may be less for a more intense laser beam. Filters may be calibrated by measuring both the incident power and the transmitted power, but the measurement must be done at low enough irradiance so that the filter (and the detector) does not become saturated.

One useful method for attenuating the beam before detection is to allow it to fall normally on a diffusely reflecting massive surface, such as a magnesium oxide block. The arrangement is shown in Figure 5.20. The goniometric distribution of the reflected light is independent of the azimuthal angle and depends on the angle q from the normal to the surface in the following simple manner:

$$P_\omega d\omega = \frac{P_{tot} \cos\theta d\omega}{\pi} \tag{5.14}$$

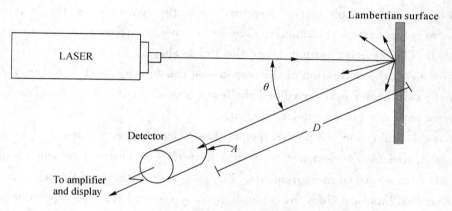

Figure 5.20 Arrangement for measuring laser power output

where P_ω is the power reflected into solid angle $d\omega$ at angle θ from the normal, and P_{tot} is the total power. This relation is called Lambert's cosine law, and a surface that follows this law is called a Lambertian surface. Many practical surfaces follow this relation approximately. The power that reaches the detector after reflection from such a surface is

$$P_{detector} = \frac{P_{tot} \cos\theta A_d}{\pi D^2} \tag{5.15}$$

where A_d is the area of the detector (or its projection on a plane perpendicular to the line from the target to the detector), and D is the distance from the reflector to the detector. This approximation is valid when D is much larger than the detector dimensions and the transverse dimension of the laser beam. With a Lambertian reflector, the power incident on the photosurface may be adjusted simply in a known way by changing the distance D. The beam may be spread over a large enough area on the Lambertian surface so that the surface is not damaged. The distance D is made large enough to ensure that the detector is not saturated. The measurement of the power received by the detector, plus some easy geometric parameters, gives the total beam power.

One widely used calibration method involves measurement of the total energy in the laser beam (with a calorimetric energy meter) at the same time that the detector response is determined. The temporal history of the energy delivery is known from the shape of the detector output. Since the power integrated over time must equal the total energy, the detector calibration is obtained in terms of laser power per unit of detector response.

In many applications, one uses a calorimeter to calibrate a detector, which is then used to monitor the laser output from one pulse to another. A small fraction of the laser beam is diverted by a beam splitter to the detector, while the remainder of the laser energy is delivered to a calibrated calorimeter. The total energy arriving at the calorimeter is determined. The detector output gives the pulse shape. Then numerical or graphical integration yields the calibration of the response of the detector relative to the calorimeter. Finally, the calorimeter is removed and the beam is used for the desired application, while the detector acts as a pulse-to-pulse monitor.

Electrical calibration of power meters has also become common. The absorbing element is heated by an electrical resistance heater. The electrical power dissipation is determined from electrical measurements. The measured response of the instrument to the known electrical input provides the calibration. It is assumed that the deposition of a given amount of energy in the absorber provides the same response, independent of whether the energy was radiant or electrical.

The difficulty of accurate measurement of radiant power on an absolute basis is well known. Different workers attempting the same measurement often obtain substantially different results. This fact emphasizes the need for care in the calibration of optical detectors.

5.5.6 Power supplies for optical detectors

The basic power supply for a photodetector consists of a bias voltage applied to the detector and a load resistor in series with it. The basic circuit for a photoconductive detector is shown in Figure 5.21. As the irradiance on the detector element changes, its conductance changes because of the free carriers generated within it. A change in the conductance increases the total current in the circuit and decreases the voltage drop across the detector. The load resistor is necessary to obtain an output signal. If the load resistor were zero, all the bias voltage would appear across the detector and there would be no signal voltage available. In the circuit shown, an increase in light intensity increases the voltage drop across the resistor, yielding a signal that may easily be monitored. If the light intensity is modulated in a periodic fashion, an AC signal will be detected.

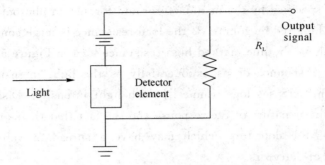

Figure 5.21 Basic circuit for operation of a photoconductive detector. The load resistor is R_L.

The magnitude of the available signal increases as the value of the load resistor increases. But this increase in available signal must be balanced against possible increase in Johnson noise and possible increase in rise time, because of the increased RC time constant of the circuit. The designer must trade these effects against each other to obtain the best result for the particular application.

A photovoltaic detector requires no bias voltage; it is a voltage generator itself. The basic circuit for a photovoltaic detector is shown in Figure 5.22. This shows the conventional symbol for a photodiode at the left. The symbol includes the arrow representing incident light. The incident light generates a voltage from the photodiode, which causes current to flow through the load resistor. The resulting IR drop across the resistor again is available as a signal to be monitored.

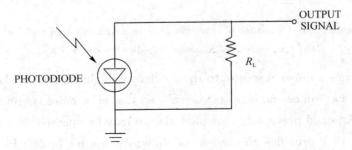

Figure 5.22 Basic circuit for operation of a photovoltaic detector. The symbol for a photodiode is indicated. The load resistor is R_L.

Optical Sensing and Measurement

In this configuration it is assumed that the value of the load resistor is much larger than the value of the shunt resistance of the detector. The shunt resistance is the resistance of the detector element in parallel with the load resistor in the circuit. The value of the shunt resistance is specified by the manufacturer and for silicon photodiodes may be a few megohms to a few hundred megohms. If the load resistance is large compared to this, the operation will be along the line marked high-resistance load in Figure 5.18. The value of the detector shunt resistance drops exponentially as the light intensity increases. The output voltage then increases logarithmically with light intensity. Disadvantages of this circuit are the nonlinear nature of the response and the fact that the signal depends on the shunt resistance of the detector, which may have a spread in values from different production batches of detectors.

To counter these disadvantages, a photovoltaic photodiode is often used in a circuit such as shown in Figure 5.23. In this case, the load resistance has a value much less than the shunt resistance of the photodiode. The operation thus corresponds to the line marked low-resistance load in Figure 5.18. Again the photocurrent flows through the load resistor and produces the observed signal. The photocurrent is fed to the virtual ground of an operational amplifier. This provides amplification to counter the decreased voltage drop resulting from the low value of the load resistor.

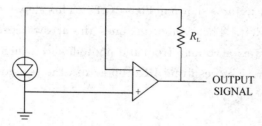

Figure 5.23 Circuit for photovoltaic operation with low load resistance. The load resistor is R_L and the feedback resistor is R_F.

This circuit has a linear response to the incident light intensity. It also is a low-noise circuit because it has almost no leakage current, so that shot noise is eliminated.

We have mentioned previously that photodiodes may be operated in a photoconductive mode. A circuit that provides this mode is shown in Figure 5.24. In this mode, the photocurrent produces a voltage across the load resistor which is in parallel with the shunt resistance of the detector. In this mode, the shunt resistance is nearly constant. The diode is reverse biased, so that the operation is in the third quadrant of Figure 5.10. One may use large values of load resistance, to obtain large values of signal, and still obtain linear variation of the output with light intensity.

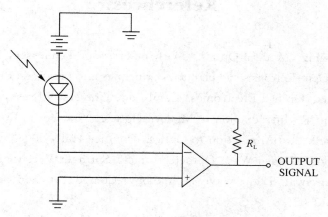

Figure 5.24 Circuit for operation of a photodiode in the photoconductive mode. The load resistor is R_L and the feedback resistor is R_F.

This circuit is capable of very high-speed response. It is possible to obtain rise times of one nanosecond or below with this type of circuit.

The biggest disadvantage of this circuit is the fact that the leakage current is relatively large, so that the shot noise is increased.

Many different types of circuits have been designed for photodetector operation for particular applications. We will present two specialized circuits for specific applications. These will give an idea of the range of circuits that may be employed for detection of radiant energy.

References

[1] Marianne Breinig. Modern Optics. The University of Tennessee, 2008.
[2] S. O. Kasap. Optoelectronics and photonics principles and practices. Prentice Hall, 2003.
[3] A. K. Ghatak. Optical Electronics. Cambridge University Press, 2007.
[4] Fowles, Grant R. Introduction to Modern Optics. New York: Dover, 1975.
[5] Pedrotti, Frank L. Introduction to Optics. Prentice Hall, 2006.
[6] Tamir Theodor. Guided-Wave Optoelectronics. Springer Verlag, 1990.
[7] Govind P. Agrawal. Light Wave Technology: Components and Devices. Wiley-Interscience, 2007.
[8] A. K. Ghatak. An Introduction to Fiber Optics. Cambridge University Press, 1998.
[9] Rosencher, Emmanuel. Optoelectronics. Cambridge University Press, 2003.
[10] Joseph W. Goodman. Introduction to Fourier Optics. Roberts and Co. Publishers, 2007.
[11] Deepak Uttamchandani. Principles of Modern Optical Systems. Artech House Publishers, 2007.
[12] Horst Zimmermann. Integrated Silicon Optoelectronics. Springer-Verlag Telos, 2007.